INNOVATION, COMMERCIALIZATION, AND START-UPS IN LIFE SCIENCES

INNOVATION, COMMERCIALIZATION, AND START-UPS IN LIFE SCIENCES

Second Edition

James F. Jordan

CRC Press is an imprint of the
Taylor & Francis Group, an informa business

Cover design by Sarah Mailhott.

Second edition published 2022
by CRC Press
2 Park Square, Milton Park, Abingdon, Oxon, OX14 4RN

and by CRC Press
6000 Broken Sound Parkway NW, Suite 300, Boca Raton, FL 33487-2742

First edition published by CRC Press 2015

CRC Press is an imprint of Informa UK Limited

British Library Cataloguing-in-Publication Data
A catalogue record for this book is available from the British Library

Library of Congress Cataloging-in-Publication Data
Names: Jordan, James F., author.
Title: Innovation, commercialization, and start- ups in life sciences /
James F. Jordan.
Description: Second edition. | Boca Raton, FL : CRC Press, 2022. | Revised
edition of the author's Innovation, commercialization, and start-ups in
life sciences, [2015] | Includes bibliographical references and index. |
Summary: "This book provides the reader the methodologies necessary to
create a successful life sciences start-up from initiation to exit.
Written by an expert who has worked with more than 400 life sciences
start-ups, the book discusses specific processes and investor milestones
that must be navigated to align customer, funder, and acquirer needs"--
Provided by publisher.
Identifiers: LCCN 2021019176 (print) | LCCN 2021019177 (ebook) | ISBN
9780367533038 (hardback) | ISBN 9780367533045 (paperback) | ISBN
9780367533052 (ebook)
Subjects: LCSH: Biotechnology industries. | New business enterprises.
Classification: LCC HD9999.B442 J67 2022 (print) | LCC HD9999.B442
(ebook) | DDC 338.4/76606--dc23
LC record available at https://lccn.loc.gov/2021019176
LC ebook record available at https://lccn.loc.gov/2021019177

ISBN: 978-0-367-53303-8 (hbk)
ISBN: 978-0-367-53304-5 (pbk)
ISBN: 978-0-367-53305-2 (ebk)

Access the Support Material: https://healthcaredata.center

DOI: 10.1201/9780367533052

Typeset in Times
by MPS Limited, Dehradun

Dedication

Dedicated to my wife Marcy, with her love my life is unshakable; without her talent this book would be unreadable; and to my children Sarah, Natalie, Danielle, and Julia, bringers of great joy and numerous surprises.

Contents

SECTION III Align With The Industry Norms

Foreword

There are an almost unlimited number of business sectors for a potential technology startup company to focus on and profit in. But there will always be a certain degree of nobility and honor to those entrepreneurs who choose life sciences. Life sciences offers a clear opportunity not only for success and profit but also to benefit all of humanity. As someone who at different points in my career has been an information technology executive, startup company COO, angel investor, adviser, consultant, and venture capitalist, I've seen and heard a lot of investment pitches. Every founder aims to portray his startup in sometimes heroic terms, but those in the life sciences ring a bit truer. They offer a reward beyond just a high investment return.

Life sciences startup companies usually fall into one of four major categories: diagnostics determining the patient's issue or disease, medical devices for treatment or remediation, or pharmaceutics. Tying all these categories together is medical information technology (IT) which harnesses the data needed to guide development, measure efficacy, quantify success, and provide care. Working together and properly integrated, these life sciences categories bring us closer and closer to the ideal and vision of a real-time health system.

In a real-time health system diagnostics, medical devices, pharmacology, and healthcare delivery all work together to maximize patient benefit, speed care, and reduce costs. The connectivity that makes a real-time health system possible is the successful integration of information technology. However, especially in the United States, it seems like information management in the healthcare sector hasn't advanced as far or as broadly as it has in other major sectors of the economy.

Over the last 150 years, really since the understanding of germ theory, there have been amazing advances across all of these life sciences categories. We've gone from the first widely available antibiotics in the 1940's to today's vast pharmacopeia. Today's diagnostic tools and medical devices look like science fiction medicine from just a few decades ago. We're the beneficiaries of stunning medical advances across a wide spectrum including surgical techniques, artificial organs, transplant technologies, gene therapies, and advanced treatment regimens.

And yet, through all of this, healthcare information technology isn't quite there yet.

Most recently, I saw first-hand how government regulation, business practices, and a lack of effective information technology impacts the U.S. health insurance industry. Each U.S. state handles health insurance rates differently. A few states make the rates public allowing insurance brokers to easily guide their clients to the best policies for their needs. Other states make only some data available or the data available doesn't include discounts or incentives offered by insurers. And in other states, little or no rate data is available at all. As you can imagine, trying to develop an information technology system to provide a simple comparison of medical insurance plans for customers is almost impossible in this environment.

You don't have to look at an industry as complex as health insurance to see the healthcare IT problem. We all experience this lack of health care information technology in our everyday lives. My doctor still faxes by phone my prescription to my pharmacy.

I can't give my new dentist any treatment information collected about me by my previous dentist over the past 20 years. My elderly parents have to manually track their many prescription drugs from different doctors and different pharmacies in a paper notebook they carry to each exam. And in the United States, we allocate a staggering 20% of our medical expenses just to pay for medical insurance administration costs. Clearly, there's an opportunity here for the next entrepreneur to revolutionize healthcare IT.

The solution for the healthcare information technology challenge, indeed for so many life sciences problems, aren't going to come from giant corporations, large healthcare providers, or bureaucratic government agencies. The solutions are waiting in the minds and life experiences of life sciences entrepreneurs and the investors who fund and advise their development. It's an exciting time – a much promised second information revolution may be at hand.

But how to begin? For while everyone knows the oft told fable of the garage-based innovator who steps heroically forward to success and fortune, the reality is quite different. This book will provide you with the knowledge to navigate the path from brilliant idea to successful startup company. Especially important is author James F. Jordan's focus on the relationship between entrepreneurs and their investors. It's not just a great idea and hard work that makes for startup success. The successful entrepreneur understands that they must identify, develop, and engage investors with their money, industry knowledge, past experience, and connections. James F. Jordan has been a deep insider in this world for the last 20 years. He brings together a triad of past medical device industry experience, venture capital leadership, and teaching expertise to give readers a guide and roadmap to the life sciences startup world. I wish you success and happiness on your journey.

Robert Sanguedolce
June 2021

Acknowledgments

When one writes one's first book acknowledgments, it may be the first time one pauses and takes a professional inventory of those who prepared him. This contemplation has resulted in a true appreciation of the saying, "it takes a community."

My life has been enhanced, both professionally and personally, by the presence of Robert Sanguedolce. His wisdom, friendship, logic, heart, and thoughtfulness are always available to me. My career started early, at 18 years old, through a cooperative education job at Raytheon Corporation. This would not have happened without the intervention of John Spanks. John set my career trajectory, mentored my early years, and most likely has no idea of the impact he made to my life. From my Raytheon experience, Donald Bernard picked up where John left off.

I entered the life sciences industry 25 years ago with C.R. Bard. To this organization and the team with which I had the pleasure to work, I have much gratitude, particularly to

- David Bradstreet, who pulled me from finance to lead me toward general management;
- Gregory Walker, who did not fire me when I let almost 200 employees leave my plant early on Christmas Eve;
- Peter Kershaw, who thought I could run engineering and subsequently made me a plant manager;
- Timothy Collins, who had my back during that journey;
- Paul LaViolette, who encouraged me to go into marketing and sales;
- Michael Kelly, one of the best people and process managers in the business; and
- Todd Schermerhorn, one of the best financial leaders I have been fortunate to work for, and with.

My first start-up experience was with Jeff Burbank, who taught me the importance of creating a value proposition that was impervious to competitive advancement. He taught me to continue to work the problem until the formula was right. He would not let the exuberance of my youth go to tactical execution until the strategy was sound. To Randall Hubbell, Tammy Harrison, and Rodney Park, who joined me on this journey and to Nancy Zeleniak, who taught me about reimbursement. To James Corbett of Boston Scientific, who taught me to set big, crazy, scary objectives. To "my Irish Betsy Rose," because her nursing background, combined with her MBA, demanded that I keep strategy aligned with clinical realities. To Alex Martin and Marcia Schallehn with whom I have been fortunate enough to work with twice in my career. Their boldness, clinical relationships, and deep understanding of their markets have allowed us to joyously take market share from our competitors. Early in my career, as a freshly minted MBA, Marcia chose not to debate my marketing strategy and instead sat me in a corner booth between our top division sales managers at a sales dinner. It is a beating that I am still recovering from today and a lesson about integrating with sales I will never forget. Consulting gives one an opportunity to observe and learn, and I thank Robert Kemp and Richard Soni for allowing me the ability to observe their gifts. To Joan Scala of MC2 Executive Search, who has not only been a career-long source of talent for me but also a mentor on changing market conditions career management. To Patrick Colletti, Lans Taylor, and Brenton Burns, board members of Net Health Systems: you changed my life. To Sean McGrail, the president of NESN, with whom I went through Boston University: your demonstration of how to sell and lead through people is only surpassed by your friendship.

This book would not be possible without the number of associations and foundations that contribute to the realization of innovation. I would like to thank the National Business Incubator Association, the Angel Capital Association, the National Venture Capital Association, the Life Sciences Greenhouse of Pennsylvania (PA), the PA Department of Community and Economic Development, the PA BIO Association, and the Coulter Foundation. Delivering innovation is a national effort and its measurement could not happen without the National Science Foundation's efforts. I am grateful to Mark Boroush of the National Science Foundation, who was always willing to take my calls and e-mails. Academia's interaction with incubators is primarily a local effort and the Cancer Prevention and Research Institute of Texas is making an impact.

The Pittsburgh region has greatly benefited by the ability of its nonprofits to work together for the benefit of the region. The Allegheny Conference on Community Development, led by Dennis Yablonsky, focuses on stimulating economic growth for the region. The Pittsburgh Technology Council's trade association collects, collates, and aligns the region's business voices to maximize impact. The Pittsburgh Venture Capital Association and the Blue Tree Angel Network, led by Kelly Szejko and Catherine Mott, respectively, ensure that funding is available for those that successfully pass through to the seed investment phase.

The Pittsburgh region's journey of innovation is frequently started in the University of Pittsburgh, Carnegie Mellon University, and Duquesne University.

The gap between universities to seed investment would not be bridged without those willing to take on pre-seed investment risk. Outside of the life sciences, Innovation Works, led by Richard Lunak with Jim Jen and Larry Miller has made a material impact on the region. As the business of life sciences is specialized, the region is fortunate to have an incubator called the Pittsburgh Life Sciences Greenhouse (PLSG) which has partnered with LifeX Labs in 2020 to help navigate its complex environment. The organization's success has been due to its employees and advisors such as Harold Safferstein, Donald Taylor, Alicia Varughese, Diana Cugliari, Debra Parrish, Paul Kornblith, Gerome Granto, and Lans Taylor. This organization's executive-in-residence program is designed to move executives into the community. There are over 50 individuals who have done so who are not included here solely due to the space required to name them all. However, I must give special thanks to Alan West and Craig Markovitz who offered specific stories included in this book. The Commonwealth of Pennsylvania, University of Pittsburgh, Carnegie Mellon University, University of Pittsburgh Medical Center, and Pittsburgh's regional foundation community founded PLSG as a public/private partnership. The PLSG would not have happened if not initially funded by the State of Pennsylvania and its foundations, specifically the McCune Foundation, the Richard King Mellon Foundation, and the Heinz Endowments.

It was my academic involvement at Carnegie Mellon University that encouraged me to pursue writing a book and my first encourager was James Osborn. I am so grateful to my first dean, Mark Wessel, for having the courage to take a risk on me at Heinz College. I went to him to discuss health care and biotechnology and was encouraged to shape their Masters in Biotechnology and Management Program. Subsequently, Dean Ramayya Krishnan encouraged me to think more broadly about health care and the topic of innovation: he continues to challenge my thoughts on this matter daily. I am grateful to interact with the talented staff and academics at Heinz College including Denise Rousseau, Rema Padman, Laura Synnott, Kristen Kurland, David Dausey, and Martin Gaynor. I thank the associate deans, Andrew Wasser and Jacqueline Speedy, for their tolerance as the health care programs horizontally cross their structures. Jacqueline helped me get approval for new Master's Program which is a complex matter in a university setting. Her relationships, political astuteness, analytical skills, and sense of humor removed so many of those obstacles. I am appreciative of Alexandra Lutz, an esteemed Heinz College alumna for returning to lead these programs after my returning to a more academic position. Continued thanks to my now retired colleagues Gerrie Halloran, Patricia Lee, and Barbara Pacella for their daily management of the chaos between attending to the personal needs of the individual students and running nationally ranked programs. The ability to simultaneously nurture the hearts and minds of master's students and also have the organized discipline to administer a program is not an easy task. I am grateful to my market research team members Cyd Charisse Mercado and Ruby Rose Pakinkin for the never ending effort to keep our research updated and available to our students via https://HealthcareData.Center.

I am grateful to the founder of Medrobotics Corporation, Howie Choset, for letting me participate in his invention from the ground floor. A special thank you to Richard Petersen, who redlined my early thoughts on this book and taught me several techniques he originated that cannot be shared here.

I thank my parents James and Sally Jordan, whose 38 years as educators are an example of giving back and a demonstration of what "a calling" means. My mother was a kindergarten teacher and when she passed in 1999, I was amazed by the number of people that showed up at her wake. There was one family where the grandmother, the mother, and the granddaughter had all had my mother as a teacher. Now that is impact! It was only later in life that Michael Termini would provide me the opportunity to teach at the University of Richmond where he helped me understand my parents' calling. My father was a middle school principal and I was fortunate to see his 40th birthday roast from his staff. When the fun was over, his staff spoke of his ability to listen and be fair and consistent in his leadership. I remember witnessing this at a young age and thinking, those are not sexy words, but can an administrator aspire to anything greater? This memory visits me frequently and I aspire daily to try to be that person. I thank my most amazing siblings, Christopher Jordan, Michael Jordan, Marc Jordan, and Jennifer McGrath, who have all tolerated my lack of presence over the years as I pursue these interests. They always welcome me with the warmest of hearts. To my children, Sarah, Natalie, Danielle, and Julia, who all have their own gifts to unfold unto the world. Since this book was first publishing, I have witnessed these women graduate and find unique ways to contribute to the world. Observing their adventures, courage, love, and desire to contribute to the world brings me great joy. Finally, to my wife, who allowed my voice to be accompanied by good grammar, I can only express my gratitude by sharing a moment when, one evening after a long day of her helping me to write and edit, I stated that she was more important to me than a wife. Her tears suggested that she is truly my partner, both in my heart and my successes. A finally to our new family puppy Nori who often drags me into the woods for a walk and recognizes that I need a reflection break even when I think I don't.

Finally, I am grateful my editor, Marc Gutierrez of Taylor & Francis, who allowed me to update this book. During our time working together, Marc has started his family too and I cannot offer enough blessings.

About the Author

James F. (Jim) Jordan is a senior executive with more than 25 years experience in global Fortune 25, entrepreneurial start-up and venture investment organizations in life sciences, health IT, and health care.

He is President and CEO of StraTactic, Inc., which enables both Fortune 500 and start-up companies to develop their go-to market, commercialization, and business development strategies. He is also a Distinguished Service Professor of Health Care and Biotechnology at Carnegie Mellon University's Heinz College, where he has published numerous articles and books on innovation, start-ups, intellectual property, and health systems.

Jim is an active industry expert in healthcare and the life sciences. With a lifelong passion for learning, Jim's business accomplishments, combined with his academic achievements, enable him to apply his broad field of knowledge and experience to his clients. A dynamic communicator, Jim is an active author, speaker, and consultant.

Previously, Jim served as President and CEO of the Pittsburgh Life Sciences Greenhouse (PLSG), a public-private economic and venture fund owned by Carnegie Mellon University and the University of Pittsburgh. Jim was at the PLSG from 2005 to 2020 and applied his more than 25 years of experience to work with over 493 life science startup companies, with direct investment in 93 of them.

Prior to joining the PLSG, Mr. Jordan served as Senior Vice President of a $3 billion division of McKesson Corporation and as a Vice President for Marketing at Johnson & Johnson. He has leveraged this experience in several startup ventures and is active on several Boards of Directors. Mr. Jordan's experience also includes consulting engagements with numerous companies such as Medtronic, Frost & Sullivan, Circuit City, Philip Morris, Northrop Grumman, Schwartz Pharmaceutical and Otsuka Pharmaceutical, among others.

Section I

Innovation Is a Process of Connected Steps

1 Investment Uses a Translation Process to Deliver Innovation

Innovation is defined as introducing something new to affect change. As it relates to our discussions on start-up companies, the conclusion of a successful innovation cycle is profit and a return for investors. This journey of innovation begins with investment, which comes from both public and private sources. Figure 1.1 depicts how investment utilizes a loosely connected translation process to deliver its innovation.

The translation process itself has micro processes and outputs embedded in it. Defining these components allows us to have an effective conversation and establish our dialogue for the remainder of this book. Let's start with the word *research*. Research is the systematic investigation into, and study of, materials and sources in order to establish facts and reach new conclusions.[1] Research can be broken down into further components such as basic research, applied research, and translational research.

Basic research has the goal of advancing or increasing the understanding of a fundamental principle. It has no immediate intent to yield a product; its intent is to advance knowledge.[2] There is probably no better example of basic research than the National Center for Human Genome Research (NCHGR), which was initially sponsored by the U.S. Department of Energy's Office of Health and Environmental Research and later joined by the National Institutes of Health (NIH). This basic research program was conducted to develop technologies for rapid geno-typing, developing markers that were easier to use, along with new mapping technologies. Investments were made in increasing DNA sequencing capacity and focusing on systems integration for better data analysis. Other investments in technology development included the modeling of organisms. Informatics investments were oriented toward technology transfer so that the results could be easily directed toward the next phase of research—applied research. The final sequence was completed in 2004 and made available on the Internet.[3]

Applied research is dedicated to answering a specific question to solve a practical problem. It accesses and uses some part of the research community's accumulated theories, knowledge, methods, and techniques. Its intent is focused on a specific academic, commercial, or business objective. A story published by Andrew Pollack in *The New York Times* on June 14, 2010, titled "Awaiting the Genome Payoff," is an example of basic research feeding applied research. This article speaks to the "cornucopia of new drugs" predicted from the genome project. Companies such as Amgen, Merck, Novartis, Bristol-Myers Squibb, and others are

DOI: 10.1201/9780367533052-1

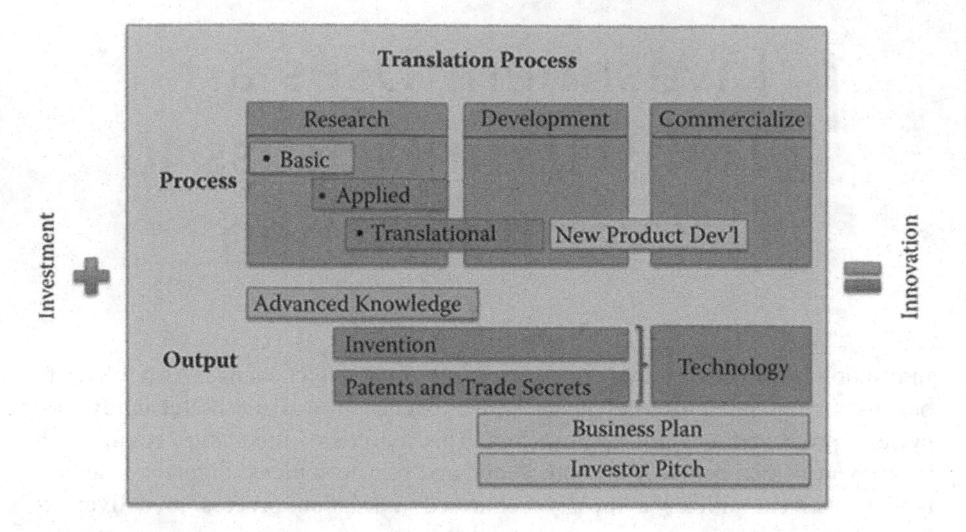

FIGURE 1.1 The investment translation process.

using the output of the genome's basic research as an input into their drug development process. This article estimated that, at the time, two-thirds of the drugs in development utilized the basic research data derived from the Human Genome Project.[4]

Translational research is a specific term used in human health and life science research efforts. Translational research utilizes multidisciplinary collaborations to make findings from basic research useful for practical applications to enhance human health. It is primarily practiced in the medical, behavioral, and social sciences.[5] According to the NIH, "Translational research includes two areas of translation. One is the process of applying discoveries generated during research in the laboratory, and in preclinical trials, to the development of trials and studies in humans. The second area of translation concerns research aimed at enhancing the adoption of best practices in the community."

Translational research is broader than applied research. Figure 1.2 helps expand our discussion. Proteomics is the study of the composition, structure, function, and interactions of proteins with living cells. Applied research could pick a particular aspect of a proteomic discovery and determine its association with a specific disease. This alone could be considered a successful chapter in applied research. The base of the pyramid in Figure 1.2, identified as *Advance Target Identification*, contains the applied research component of our proteomics discussion. In a non-translational research environment, this is where the applied discovery chapter may stop. However, in a translational research environment whose boundaries can be as broad as those identified by the dotted box, a multidisciplinary team may be formed to answer the broader question of how this discovery could be made into a drug and tested clinically.

As an example of the grayness of the translational process, we can look to Carnegie Mellon University's (CMU) Robotics Institute for two examples. Branislav Jaramaz,

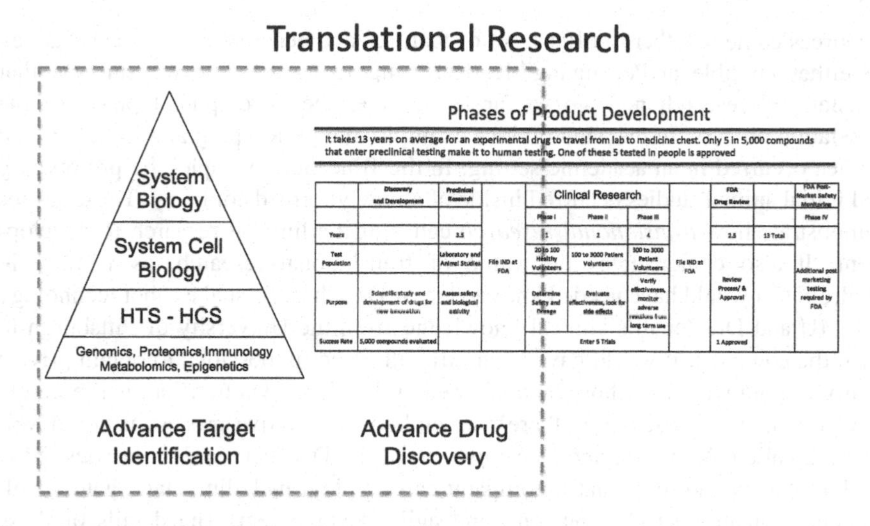

FIGURE 1.2 Translational research. (By Dale E. Wierenga and C. Robert Eaton, Office of Research and Development, Pharmaceutical Manufacturers Association.) *Source: Innovation, Commercialization and Start-ups in Life Sciences*, by James F. Jordan, CRC Publishing 2014.

PhD. was a member of the technology team at the university that developed an orthopedic product. As the company's target market was clearly identified, Dr. Jaramaz, partnered with a local entrepreneur, Craig Markovitz, to license the technology into a new company Blue Belt Technologies. They were also joined by Dr. Anthony DiGioia, an internationally recognized pioneer in orthopedics. The team was combination of what translation requires; a technologist (Dr. Jaramaz), a physician (Dr. DiGioia, and a business leader (Mr. Markovitz). This translational research resulted in a $275m Acquisition by Smith & Nephew.[6] Smith & Nephew is a $5.2 billion multinational medical equipment manufacturing company.[7]

A company called *Medrobotics Corporation* is another successful example of translational research of a medical device. Motivated by researching applications in confined spaces, Dr. Howie Choset of Carnegie Mellon University's (CMU) Robotics Institute built a research program around snake robots. Dr. Choset collaborated with Dr. Marco Zenati from the University of Pittsburgh, who is an internationally recognized pioneer in minimally invasive cardiac surgery and medical robotics. As the technology has numerous potential markets, it stayed within the university longer. The team received applied research funding to investigate the use of snake robotics in surgery. This resulted in numerous patents, research grants, and a proof of concept via animal studies. This translational research resulted in a university spinout forming the Medrobotics Corporation. The story as to this company's ultimate success has still to be written, as we will learn later in this book, there are numerous steps beyond the translational phase.

Our next process is development: the goal of the development phase is to cause something to grow or mature and advance. This is where the business and technical resources come together to offer a set of features and benefits in a product that can be either tangible and/or intangible. Referring to Figure 1.1, we can see that translational research may go so far as to enter the development phase of the translation process. In the Medrobotics example, the prototyping and initial animal studies occurred in an academic setting. In the Blue Belt example, the prototyping and initial animal studies occurred inside of a newly formed company. These stories demonstrate how *translational research* can blur the lines of research to development. It also demonstrates the value of translational research, as without it, Medrobotics would have needed knowledge of Dr. Choset's snake robot technology at CMU and Dr. Zenati's clinical knowledge from the University of Pittsburgh to form the company. It would have been difficult, if not impossible, to obtain private equity to fund what Dr. Choset and Dr. Zenati developed via translational research.

Continuing our discussion, there's an embedded iterative process inside development called *New Product Development* or NPD. There are numerous NPD methodologies and tools that a company can employ, including stage gate development, lean product development, and agile development. The details of these approaches are beyond our discussion here. However, all of the methodologies share a philosophy that require frequent pauses to ensure that the technical, business, and customer requirements maintain alignment. These processes also integrate with a company's project management system, which is used to efficiently initiate, plan, execute, monitor, and control the project.[8]

The NPD process crosses from development into commercialization and concludes with the product launch. Commercialization is the process or cycle of introducing a new product or production method to the market. The end result of commercialization is the availability of the innovation to be exploited for profit. The concept of profit is broader in a start-up company than in a publicly traded company. A publicly traded company's profit drives its stock price and investors have the capability to sell their stock. Start-ups must either take a company public (called an *initial public offering* or IPO) or sell the company to obtain liquidity for their investors. Thus in a start-up, discussions of successful commercialization frequently include the liquidity event (IPO or sale). Referring back to Figure 1.1, we have defined the process components of research, development, and commercialization. The next discussion will focus on the major outputs of the system.

The output of basic research is advanced knowledge for its own sake. For example, the American Association for the Advancement of Science (AAAS), an international nonprofit organization, gives out an annual Breakthrough Award via its magazine *Science*.[9] In 2012, the detection of the Higgs Boson, a subatomic particle, won the top scientific achievement award.[10] This basic research has no immediate commercial value; advancing the knowledge of physics motivated its pursuit.

The outputs of applied research and translational research are invention, patents, and trade secrets. To invent is to produce something for the first time and it typically manifests itself as a process or device. When something is produced for the first time, it may result in published research or a patent. A patent is the establishment of a proprietary right by a government authority, conferring exclusive use and excluding others from making, using, or selling the invention.[11] When something is produced for the first time, it may become a trade secret, which is any confidential information that provides a company with a competitive edge. A trade secret is generally in the manufacturing of the product but can include sales, distribution methods, and consumer profiling methods. Trade secrets are not a registrable right like a patent.

As previously stated, not all inventions are made into patents: an example may be illustrative. Continuing with the 2012 *Science* Breakthrough Awards, a recognized pioneering achievement was the discovery of a new high-coverage DNA sequencing reconstruction technique that bound special molecules to single strands of DNA to allow researchers to sequence the Denisovan genome from fragments of bone from an ancient finger.[12] Dr. Svante Paabo, and his colleagues at the Max Planck Institute of Evolutionary Anthropology in Leipzig Germany, invented this technology.[13] Their motivation for this invention was to support their research to uncover the origins of a unique species of hominids, not specifically to create a patent for future commercial gain. Juxtapose the patent filed by W.R. Grace & Company on the species of transgenic soybeans. This applied research was intended for commercial gain, as the farm value in the United States alone in 1992 was $11.8 billion.[14] A trade secret is a more elusive discussion, as its details are not publicly disclosed by intent. Perhaps the best known example of a trade secret is the formula for Coca-Cola. This formula is commercially valuable, yet not protected by a government registration, but rather protected through an internal process by the company. Invention, patents, and trade secrets are all outputs of applied and translational research. Are these outputs the inputs into something else?

Start-up investors frequently speak to investing in technology. *Google Dictionary* defines technology as the scientific application of knowledge for practical purposes in a particular industry. For the purposes of this book, invention, patents, and trade secrets, the outputs of research, are to be considered the inputs to the use of the word *technology*.

DISCUSSION QUESTIONS:

1. What are the three major process components of research?
2. Define and discuss the investment translation process?
3. What are the differing motivations and outcomes between the research process and the investment *translation* process?
4. Give an example on why translational research is broader than applied research?
5. What is the phase of a pharmaceutical clinical trial?
6. What is the result or goal of commercialization?

NOTES

1 *Google Dictionary*, s.v. "Research," https://www.google.com/search?q=what+is+research&oq=what+is+research&aqs=chrome.69i57j0j69i60j0l2j69i60.2782j0j7&sourceid=chrome&espv=210&es_sm=119&ie=UTF-8 (accessed March 15, 2014).

2 *Wikipedia*, s.v. "Basic research," http://en.wikipedia.org/wiki/Basic_research (accessed March 15, 2014).
3 About NHGRI: A Brief History and Timeline, *National Human Genome Research Institute*, http://www.genome.gov/10001763.
4 Andrew W. Pollack, Awaiting the Genome Payoff, *The New York Times*, June 14, 2010.
5 What Is Translational Research? *Center for Clinical and Translational Sciences*, http://ccts.uth.tmc.edu/what-is-translational-research (accessed March 15, 2014).
6 Brad Perriello, Smith & Nephew closes $275M Blue Belt Buy, +Mass Device, January 5, 2016, https://www.massdevice.com/smith-nephew-closes-275m-blue-belt-buy/
7 At-A-Glance, Smith & Nephew company Website, February 2021, https://www.smith-nephew.com/about-us/who-we-are/at-a-glance/
8 What Is Project Management? *Project Management Institute*, http://www.pmi.org/About-Us/About-Us-What-is-Project-Management.aspx (accessed March 15, 2014).
9 *Science*, http://www.sciencemag.org/magazine (accessed March 15, 2014).
10 Brandon Bryn, Science: Detection of the Higgs Boson Is the Top Scientific Achievement of 2012, The Denisovan Genome, *The American Association for the Advancement of Sciences*, December 20, 2012, http://www.aaas.org/news/releases/2012/1220sp_boy.shtml.
11 *Dictionary.com*, s.v. "Patent," s.v. "Invent" (accessed March 15, 2014).
12 Brandon Bryn, Science: Detection of the Higgs Boson Is the Top Scientific Achievement of 2012, The Denisovan Genome, *The American Association for the Advancement of Sciences*, December 20, 2012, http://www.aaas.org/news/releases/2012/1220sp_boy.shtml.
13 Katherine Harmon, New DNA Analysis Shows Ancient Humans Interbred with Denisovans, *Nature: Journal of Science*, August 31, 2012, http://www.nature.com/news/new-dna-analysis-shows-ancient-humans-interbred-with-denisovans-1.11331.
14 Species Patent on Transgenic Soybeans Granted to Transnational Chemical Giant W.R. Grace, *ETC Group*, March 30, 1994, http://www.etcgroup.org/fr/node/489.

2 Investment Is Critical to a Nation's Prosperity

Investment is critical to a nation's prosperity. To further this discussion, one must understand how nations measure their economic value. A nation's economic value is calculated by measuring all of its investments and expenditures on goods and services. This economic value is called *gross domestic product* (GDP) and is calculated by adding total consumption, investment, government spending, changes in inventories, and net exports.[1]

GDP is a good measure of a nation's economic progress and national standard of living as it is adjusted for inflation. Inflation occurs when the cost of a specific good or service increases. This increase is either due to an increase in demand, given a limited supply of a product, or an increase in the nation's money supply.[2] Excluding inflation in GDP comparisons allows true comparison of economic progress; to appreciate the benefit of comparison a sample of countries' 2018 GDP is provided[3] in Table 2.1.

Merriam Webster defines wealth as an abundance of valuable material possessions or resources. The intention of investment is to place human capital, an asset, or cash to work in the hope that it will appreciate in the future. Appreciation (or growth) of the investment generates wealth, which can either be spent in the future or continued as investment. To increase a nation's standard of living, countries seek to perpetually increase their GDP.[4]

Gross Investment is the only GDP component intended to generate long-term wealth.[5] Please note the difference between the terms *Gross Investment* and *Investment*. This is to acknowledge that there are components of government spending on research and development that are oriented toward creating long-term wealth. Additionally, some inventories such as oil and forestry reserves are held with the intention of long-term appreciation. With the exception of gross investment, all the other components of the GDP formula are consumed and spent in the now. This subtlety is necessary to differentiate between the economist's definition of investment in the GDP formula and the sources of investment that translate into the delivery of innovation. For simplicity, the term *investment* is used through the balance of this economic discussion with the meaning of *gross investment.*

DISCUSSION QUESTIONS:

1. How do you calculate a nation's Gross Domestic Product (GDP)?
2. What is the main intention of investment?
3. Why is GDP a good measure of a nation's economic progress?

DOI: 10.1201/9780367533052-2

TABLE 2.1
Global 2018 GDP Comparative

Country	GDP in Dollars (In Trillions)	% GDP Growth
United States	20.5T	2.90
China	13.6T	6.60
Japan	5T	0.80
Germany	3.9T	1.50
United Kingdom	2.9T	1.40
France	2.8T	1.70
India	2.7T	6.80
Italy	2.1T	0.80
Brazil	1.9T	1.30
Canada	1.7T	1.90

Source: The World Bank, https://data.worldbank.org/indicator/NY.GDP.MKTP.CD?end=2018&start=1960

NOTES

1 Economics A-Z terms, *The Economist*, http://www.economist.com/economics-a-to-z/g#node-21529906 (accessed March 15, 2014).
2 Economics A–Z terms, *The Economist*, http://www.economist.com/economics-a-to-z/i#node-21529397 (accessed March 15, 2014).
3 Gross Domestic Product Table by Country, *Trading Economics*, http://www.tradingeconomics.com/ (accessed March 15, 2014).
4 Note: Purchasing power is the real goal of GDP. Adjustments for inflation are already made and population growth must be made to get to a per capita number.
5 It can be argued that in some industries such as oil and forestry that inventories can be considered a generator of wealth as inventories do not expire or become outdated if held as an investment.

3 The Journey of Innovation Begins with Investment

Our previous discussion allowed for the comparison of the U.S. economy, as defined by GDP, with that of other nations. As the journey of innovation begins with investment from both public and private sources, our discussion turns to the percentage of GDP spent on research & development (R&D).

Timing and global comparison are the biggest data challenges for R&D, as the pathway from federal spending through R&D requires integrating corporate data that is typically based upon a calendar year format (January to December). Multinational companies have both national and international components that must be teased out of their financial and tax reports. Adding to the complexity, federal and state government data is in a fiscal year format (July to June). Simply combining data sets is not possible. From a federal government perspective, data sets are also difficult, as the data formats differ between proposed budgets, appropriated budgets, and actual spending.

The National Science Board publishes the *Science and Engineering Indicators* (SEI) derived from these vast sources, and as such, data lags by a few years. For example, the 2020 SEI are based upon 2017 data.[1] Given the complexity and timing of the data sets, it is important to trend the data to uncover its true story.

As it relates to our discussion, the question is: how much investment in R&D is placed into the U.S. economy annually? Since 1990, the U.S. economy has invested roughly 2.5% of its GDP into research and development. Table 3.1 unfolds our most recent history as increases from 2012 to 2018 represent an increase to 2.8%. As the main interest of this book is life sciences, it should be noted that the SEI for the calendar year 2017 reports that 4.8% of all U.S. business R&D is performed in the pharmaceuticals and medicines group."[2]

The next question: how does the U.S. R&D spending compare to other nations, both in terms of invested dollars and as a percentage of GDP? Table 3.2 is an excerpt from the most recent science and engineering (SEI) indicators.[3]

The United States, China, and Japan represent the top three global investors in R&D. Although the United States continues to outperform the rest of the world, the SEI reports that U.S. leadership is declining. The United States accounted for 38% of global R&D in 1999, in 2017 that number decreased to 26%. Asia (excluding Russia) represented 42% of global R&D in 2017. The SEI reports "China's overall R&D investment has been growing at about 20% for over a decade."[4]

DOI: 10.1201/9780367533052-3

TABLE 3.1
2018 Science and Engineering Indicators—Current Dollars

	Sources of R&D Funding (*)					
Year	Business	Gov't	Other (***)	Total	U.S. GDP (**)	% Investment to GDP
2012	302.3	52.1	79.2	433.6	16,197.0	2.7
2013	322.5	51.1	80.4	454.0	16,784.9	2.7
2014	340.7	52.7	82.0	475.4	17,527.3	2.7
2015	355.8	52.8	85.0	493.7	18,224.8	2.7
2016	374.7	51.2	89.8	515.6	18,715.0	2.8
2017	400.1	52.6	95.2	547.9	19,519.4	2.8
2018	422.1	58.2	99.7	580.0	20,580.2	2.8

Source: * NCSES, National Patterns of R&D Resource, NSF 20–307, January 08, 2020, Table 2** GDP, U.S. Bureau of Economic Analysis, Current-Dollars*** Other includes higher education and non-profits

TABLE 3.2
R&D Expenditures as a Percentage of GDP

Region/Country/Economy	R&D Expenditure (PPP $ billions)	R&D/GDP (%)
North America		
United States (2017)	548.984	2.81
Canada (2017)	27,162	1.59
Mexico (2016)	11,259	0.49
Europe		
EU (2017)	430,121	1.97
Germany (2017)	132.004	3.04
France (2017)	64,672	2.19
United Kingdom (2017)	49,345	1.66
Italy (2017)	33,542	1.35
Spain (2017)	21,932	1.21
Asia (Central, South, East)		
China (2017)	495,980	2.15
Japan (2017)	170,900	3.20
South Korea (2017)	90,979	4.55
India (2015)	49,746	0.62
Russia (2017)	41,868	1.11
Taiwan (2017)	39,296	3.30
Pakistan (2017)	2,569	0.24
Australia (2015)	21,151	1.88

Source: Science & Engineering Indicators, Table 4–5 https://ncses.nsf.gov/pubs/nsb20203/cross-national-comparisons-of-r-d-performance#tableCtr1875

Returning our discussion to the United States, Table 3.2 indicates that, in 2017, 2.8% of GDP was invested in R&D. Our next exploration will answer the following questions:

- What was the size of the U.S. GDP in 2017?
- Of the total R&D budget, how much was spent on basic research, applied research, and development?
- What were the sources of this funding?
- Who performed the basic research, applied research, and development activities?

Figure 3.1 details the flow of R&D monies from U.S. Gross Domestic Product through to development monies for calendar year 2017. Take a moment to validate how Figure 3.1 links to Table 3.2, comparing U.S. R&D spending to the rest of the world. Figure 3.1 continues our analysis of teasing apart the components of R&D spending.

Let's walk through Figure 3.1, starting with the column titled *R&D Funding*. In 2017, $547,886 billion was spent on R&D, which was 2.8% of the GDP. There are two columns underneath this heading titled *Source* and *Performer*. *Source* is defined as the group that provided the money and *Performer* is defined as the group that actually performed the work. This delineation is important, as government funding for R&D rarely requires repayment to the government. This necessitates another pause because many might be asking the following question at this point: "Why is the author getting into all this detail in a book about start-ups?" The answer is non-dilutive funding.

Funding a start-up requires capital. To get profit back to investors the company must create a liquidity event such as a sale or IPO. When a company is sold, the difference between the selling price of the company and the capital put into the company is the return on the investment to investors. What if some of the capital put into the company came from the government? Let's assume a company sold for $100,000 and the capital put into the company was $60,000. Taking the sale price minus the capital of $60,000 would leave a return to investors of $40,000. However, what if the capital of $60,000 put into the company included a $20,000 grant from the government? Investors would only be required to invest $40,000 ($60,000 – $20,000). Now the company's return on the sale to investors would look very different if they received this grant.

Since the government does not require repayment on the $20,000 grant, it is called *non-dilutive*, as it does not dilute an investor's return. The investors get an additional $20,000 or a 20% increase in their return over what they would have received without the grant. However, our pause should demonstrate the value of continuing our discussion.

Continuing our investigation of Figure 3.1, let's review the details of the heading *R&D Funding* with the columns of *Source* and *Performer*. Under the column *Source*, the government funded $120,961 million of R&D in 2017. Underneath the $120,961 million, it is calculated that this is 22% of R&D funding (120,961/547,886). Contrast this with the *Performer* column, where the government actually

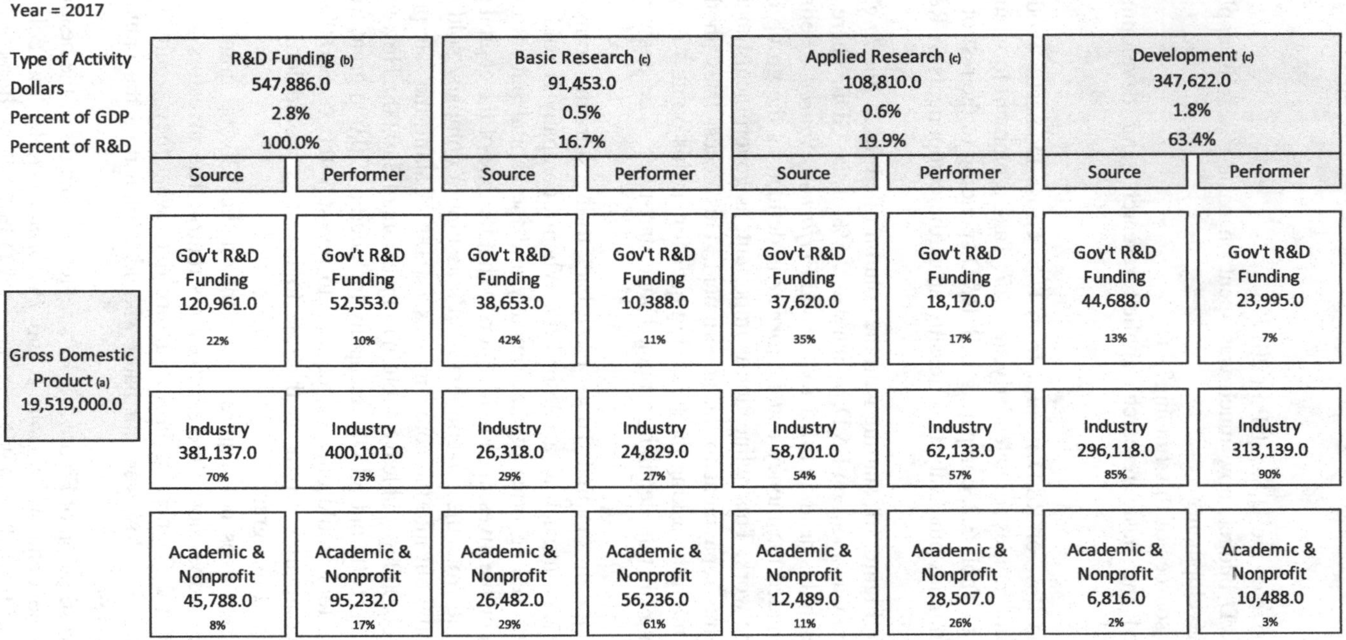

Year = 2017

Type of Activity	R&D Funding (b)		Basic Research (c)		Applied Research (c)		Development (c)	
Dollars	547,886.0		91,453.0		108,810.0		347,622.0	
Percent of GDP	2.8%		0.5%		0.6%		1.8%	
Percent of R&D	100.0%		16.7%		19.9%		63.4%	
	Source	Performer	Source	Performer	Source	Performer	Source	Performer
Gov't R&D Funding	120,961.0 22%	52,553.0 10%	38,653.0 42%	10,388.0 11%	37,620.0 35%	18,170.0 17%	44,688.0 13%	23,995.0 7%
Industry	381,137.0 70%	400,101.0 73%	26,318.0 29%	24,829.0 27%	58,701.0 54%	62,133.0 57%	296,118.0 85%	313,139.0 90%
Academic & Nonprofit	45,788.0 8%	95,232.0 17%	26,482.0 29%	56,236.0 61%	12,489.0 11%	28,507.0 26%	6,816.0 2%	10,488.0 3%

Gross Domestic Product (a) 19,519,000.0

FIGURE 3.1 GDP to development funding. (a) https://www.bea.gov/news/2018/gross-domestic-product-4th-quarter-and-annual-2017-third-estimate-corporate-profits-4th; (b) Science & Engineering Indicators 2017, NSF 20-309, U.S. R&D in 2017, January 2020, Table 1; (c) Science & Engineering Indicators 2017, NSF 20-309, U.S. R&D in 2017, January 2020, Table 3.

performed $52,553 million or 10% of the R&D. So, if the government funded $120,961 million of R&D and only performed $52,553 million in R&D, who did the R&D? The difference of $68,408 million was performed by either industry, academia, or other nonprofits. When we look at the difference between *Source* and *Performer*, you can understand that the government invested $68,408 million into industry, academia, or other nonprofits in 2017. Our analysis continues as we investigate industry R&D. Industry is the source of $381,137 million or 70% of all R&D, however, it received $18,964 million (400,101 – 381,137) from the government and thus performs $400,101 million in R&D. Academia and nonprofits fund $45,788 million in R&D but receive $49,444 million (95,232 – 45,788) from the government and thus perform $95,232 million in R&D. Now that we have detailed the workings of the rows under the *R&D Funding* heading, the rows underneath the other headings follow the same mathematical calculations.

Continuing to the heading *Basic Research* in Figure 3.1, we learn that $91,453 million or 16.7% of R&D funding goes to basic research. As basic research has the goal of advancing or increasing fundamental knowledge, it is of little surprise that 42% of funding comes from the government and that 61% of basic research is performed by academia and nonprofits.

Applied Research is 19.9% of all R&D—only 0.6% of U.S. GDP. Applied research is dedicated to answering a specific question to solve a practical problem. It is important to note that it may take many completed cycles of applied research (which we will refer to as *chapters*) to result in a new product or technology. Looking at the rows underneath the *Applied Research* heading, one observes that academia and nonprofits are only 11% of the source of applied research funding. The remaining is split between the government and industry. However, academia is the *performer* of 26% of applied research. How can this be, as one would think that basic research should be most of academia's interest?

Prior to December 12, 1980, intellectual property arising from federally funded research became the property of the federal government. After this date, the Bayh–Dole Act created a uniform patent policy that enabled small businesses, nonprofit organizations, and universities to retain title to inventions made under federally funded research programs.[5] This act allowed academia to create a revenue stream for itself, as it could either license its technology or create a start-up company. In many cases, to prepare the technology to move from academia to industry required additional applied research funding. However, not all applied research in academia is destined for industry; some of it is needed for continued basic research. Our example in Chapter 1 of Dr. Svante Paabo is a good illustration. Dr. Paabo's interest was basic research in anthropology, but he needed to invent the technology to sequence the Denisovan genome to continue his research. The applied research of Denisovan genome sequencing was a response to a basic research obstacle.

The last heading in Figure 3.1 is *Development*, valued at $347,622 million or 1.8% of the GDP. Development accounted for 63.4% of all R&D monies in 2017 and those monies were intended to deliver a specific technology.

Let's pause as we have introduced the world technology into this discussion. Please refer to Chapter 1, Figure 1.1 and bring your attention to the *Output* section. The outputs of R&D are invention, patents, and trade secrets and these form technology. Technology is a set of features and benefits offered in a product. It is important to understand that more than one technology may be required to deliver a product. This discussion could benefit from an example.

A heart attack is a result of coronary artery disease in which plaque builds up over time and blocks one or more arteries supplying oxygen to the heart. One solution for repairing or avoiding a heart attack is to open the vessel. Frequently this is done with a coronary stent, which is a tiny wire mesh tube. The specific design of the wire mesh is a technology. The material that the stent is made out of, frequently stainless steel or nitinol, are also both technologies. A catheter that is entered into the heart delivers these stents. Catheters have many different technologies as different materials and configurations provide different benefits. Avoiding the need of going through every individual product associated with a stent procedure, one can appreciate that it can take many different technologies to deliver the solution of a stent. Referring to Figure 1.1, the New Product Development (NPD) process of creating a new stenting product would most likely result in the joining of numerous technologies. These numerous technologies would be inputs into the NPD process and go through the company's *Commercialization* process.

DISCUSSION QUESTIONS:

1. Why is global comparison considered one of the biggest data challenges for R&D?
2. Explain non-dilutive funding.
3. Break down (including percentages) the type of activities performed as it relates to funding Research and Development in 2009.
4. What is the significance of the Bayh-Dole Act?
5. What are the outputs of R&D?

NOTES

1 National Science Board, Chapter 4, Science and Engineering Indicators 2020, Table 2, https://ncses.nsf.gov/pubs/nsf20307/#data-tables (date accessed February 2021).
2 S&E Indicators 2019, National Center for Science and Engineering Statistics, National Science Foundation, and U.S. Census Bureau, Business R&D and Innovation Survey and Business Research and Development Survey, Raymond M. Wolfe, https://www.nsf.gov/statistics/2019/nsf19326/overview.htm (date accessed February 2021).
3 Source: Science & Engineering Indicators, Table 4–5, Science and Engineering Indicators 2020, https://ncses.nsf.gov/pubs/nsb20203/cross-national-comparisons-of-r-d-performance#tableCtr1875 (accessed February 2021).
4 National Science Board, Chapter 4 commentary, *Science and Engineering Indicators 2012*, http://www.nsf.gov/statistics/seind12/c4/c4h.htm#s8 (accessed March 15, 2014).
5 *30 Years of Bayh–Dole*, http://b-d30.org/ (accessed March 15, 2014).

4 The United States Helps Small Companies Conduct R&D

As previously discussed in Chapter 3, the Bayh–Dole Act enabled small businesses, nonprofit organizations, and universities to retain title to federally funded inventions.

For universities, this allows for the opportunity to license technology to governments and both large and small businesses. The mechanism to do this is called a *licensing agreement*. A licensing agreement is a written contract in which the owner of the intellectual property (IP) allows the licensee to use the IP either exclusively or nonexclusively. In return for this right, the licensee pays royalties or some other exchange.[1] A royalty is a usage-based payment methodology in which payments could be received as a percentage of a sales price or a fixed amount per unit. Straight royalties are a good mechanism to use with governments or large businesses as they have existing funding sources. As is related to start-up businesses, universities frequently ask for a stake of ownership (called *equity*) or a combination of equity and royalties for a license.

The Bayh–Dole Act created a revenue mechanism for universities and it would be natural for universities to prefer to license their technologies to governments and big business, given their more secure capital structure. However, the universities had a conundrum, as big business tends to invest into its existing markets. They prefer to outsource new markets or riskier innovations to start-ups. What does outsourcing riskier innovations to start-ups mean? It means that a big company would prefer to acquire or license from a start-up company after it has demonstrated a market's value as opposed to risking an investment in an unproven or undeveloped technology or market.

Subsequent to the Bayh–Dole Act, in 1982, the Small Business Innovation Research (SBIR) grant program was established and was followed in 1992 by the Small Business Technology Transfer (STTR) program. The objective of both of these programs is to help small U.S. companies that have commercial or public benefits conduct R&D.[2] The STTR has a unique feature in that it requires small businesses to collaborate with research institutions.[3]

Both programs have two phases of funding. The objective of Phase I is to establish the technical merit, feasibility, and commercial potential for the project. Phase I demonstration allows the company to progress to Phase II funding. Since the first edition of this book, a Phase III program for small businesses to continue to receive non-competitive funding from federal agencies has been developed. These are statistically rare and are awarded by a specific federal agency for products and

DOI: 10.1201/9780367533052-4

processes that support their mission. All SBIR/STTR's have a specific Program Officer assigned from the specific agency. It would be important during a start-ups Phase II award to start working with your Program Officer to determine if there is an opportunity for a Phase III award. Since these awards are non-competitive, an expression of interest has a high probability of success. The funding amounts and durations are detailed in Table 4.1.

For the SBIR program, federal agencies with extramural research and development budgets over $100 million set aside 3.2% (FY 2017) for SBIR grants for small companies. Currently, eleven agencies participate in the SBIR program (Table 4.2).

For the STTR program, federal agencies with extramural budgets over $1 billion are required to set aside .45% (FY 2017) for STTR grants. Currently, five agencies participate in the STTR program (Table 4.3).

The SBIR/STTR programs can be used to seed the life of a new start-up by providing non-dilutive, initial funding to demonstrate a proof of concept for private

TABLE 4.1
Grant Amounts

SBIR Program	STTR Program
Phase I: $150,000 total cost for 6 months	Phase I: $150,000 total cost for 12 months
Phase II: $1,000,000 total cost for 2 years	Phase II: $1,000,000 total cost for 2 years
Phase III: Commercial (see note below)	Phase III: Commercial (see note below)

Source: Office of Extramural Research, Grants & Funding, National Institutes of Health, https://grants.nih.gov/grants/funding/sbirsttr_programs.htm (updated January 3, 2021).

TABLE 4.2
Participating SBIR Agencies

Health & Human Services (DHHS)	Agriculture (USDA)	Commerce (DOC)
Defense (DOD)	Education (DoED)	Energy (DOE)
Homeland Security (DHS)	Transportation (DOT)	Environmental Protection (EPA)
National Aeronautics and Space Administration (NASA)	National Science Foundation (NSF)	

Source: National Institutes of Health, SBIR Participating Agencies, https://www.sbir.gov/agencies-landing (accessed February 2021).

TABLE 4.3
Participating STTR Agencies

Health & Human Services (DHHS)	**National Science Foundation (NSF)**	**Energy (DOE)**
Defense (DOD)	National Aeronautics and Space Administration (NASA)	

Source: National Institutes of Health, SBIR Participating Agencies, https://www.sbir.gov/agencies-landing (accessed February 2021).

equity investors. When the author brings this point up to start-up entrepreneurs, they frequently comment that it takes too long to get that money. They feel they can raise that money themselves in a shorter period of time. Sadly, this is frequently not the case. Future chapters will expand on the many other benefits of non-dilutive funding to the start-up.

Referring to Figure 4.1, we can see the details of the obligations each department committed to both the SBIR and STTR programs in the fiscal year 2018. One can see that the departments of HHS and DOD provide much of the spending.

We may also want to understand how SBIR/STTR funding relates to the 2017 R&D illustration in Figure 3.1. This is not a perfect calculation due to

In Millions FY'18	SBIR Program		STTR Program		All Programs	
	Dollars	%	Dollars	%	Dollars	%
DOD	$ 1,166	42.0%	$ 175	45.9%	$ 1,342	42.5%
HHS	$ 931	33.5%	$ 132	34.5%	$ 1,063	33.6%
DOE	$ 249	9.0%	$ 32	8.4%	$ 282	8.9%
NSF	$ 185	6.7%	$ 19	4.9%	$ 203	6.4%
NASA	$ 166	6.0%	$ 24	6.3%	$ 190	6.0%
USDA	$ 29	1.1%	$ -	0.0%	$ 29	0.9%
DHS	$ 17	0.6%	$ -	0.0%	$ 17	0.5%
DOC	$ 15	0.5%	$ -	0.0%	$ 15	0.5%
ED	$ 8	0.3%	$ -	0.0%	$ 8	0.3%
DOT	$ 5	0.2%	$ -	0.0%	$ 5	0.2%
EPA	$ 4	0.1%	$ -	0.0%	$ 4	0.1%
Total	$ 2,776	100.0%	$ 382	100.0%	$ 3,159	100.0%

FIGURE 4.1 SBIR & STTR 2018 Obligations. *Source:* https://www.sbir.gov/sites/default/files/SBA_SBIR_STTR_FY18_Annual_Report-Final_508.pdf.

the nature of the data collection timeframes. Gross Domestic Product and R&D data are calculated on a calendar year format (January to December). Federal budgets, which include SBIR and STTR data, are tracked according to the government's fiscal year (July to June). Appreciating this overlap, we can get a directional perspective for continued insight. Figure 4.2 has been created to provide the perspective of how SBIR/STTRs relate to total GDP, total R&D, and Total Development spending. In short, we want a sense of how much of these monies support start-up companies.

Although there are other government programs that support business, for example, some of the programs used during COVID to support business payrolls, SBIR/STTR programs are focused solely on small companies and support development. Figure 4.2 hints to the relatively small amounts going to developing start-up technology. The funding is less than 0.02% of the total economy, and less than 1% of R&D and development funding.

The SBIR/STTR is a significant tool for the entrepreneur that frequently is not used to its highest potential by start-up companies. We previously discussed in this chapter that it can take numerous technology chapters to make a product. An example was given in our discussion of the cardiac stenting procedure. This procedure requires several technologies to complete it, the stent design, the stent material, the delivery catheter, and so on. Each of these technologies could be a viable SBIR/STTR submission. Combining both Phase I and Phase II, each technology has the potential of getting $1.1 million in development support. In our example of the stent procedure, we discussed in this chapter three technologies: imagine if each technology got funded. That could be $3.3 million in funding for the start-up.

In Millions	Total $	SBIR/STTR $	SBIR/STTR %
Total GDP	$ 19,519,000	$ 3,159	0.02%
R&D	$ 547,886	$ 3,159	0.58%
Development	$ 347,622	$ 3,159	0.91%

FIGURE 4.2 SBIR/STTR Obligations as a percentage of GDP, Total R&D and Development.*Source:* Calculations derived from Figures 3.1 and 4.1.

DISCUSSION QUESTIONS:

1. What is a licensing agreement?
2. How is the Bayh-Dole Act beneficial to universities?
3. Differentiate the SBIR program from the STTR program.
4. What do the SBIR and STTR programs provide?

NOTES

1 *Business Dictionary.com*, s.v. "Licensing Agreement," http://www.businessdictionary.com/definition/licensing-agreement.html (accessed March 15, 2014).
2 *SBIR/STTR*, http://www.sbir.gov/about/about-sttr#six (accessed March 15, 2014).
3 Office of Extramural Research, Grants & Funding, *National Institutes of Health*, updated March 6, 2013, http://grants.nih.gov/grants/Funding/sbirsttr_programs.htm (accessed March 15, 2014).

5 Commercialization Is Primarily Executed through Two Organization Types

Commercialization is a process of connected steps that serve to bring a product to market. Progressive commercialization techniques embrace integration, concurrence, and/or overlap with the development process to ensure proper downstream execution.

Commercialization is executed primarily through two organizational forms: corporations and start-ups. Although there are numerous commercialization processes and philosophies, they all contain a common series of steps and reviews often referred to as *gates*.

Figure 5.1 illustrates one such type of process. In this example, there are five major components to this stage-gate process. The first is idea generation and strategic alignment, followed by opportunity assessment, then product evaluation, development, and finally commercialization itself.

The first phase of *Idea Generation and Strategic Alignment* is focused on screening the various development ideas and scoping the objectives and deliverables for the project. In the case of the corporation, ideas can come from market research, customer complaints, the sales force, the business development organization, or by customer/supplier visitation.

Let's take a slight tangent to point out that the business development organization in a corporation can be a major contributor to this process. Business development is constantly scanning the marketplace for new IP or start-up companies to acquire. The business development organization's role is to uncover and quantify return opportunities to senior management. If senior management determines that the opportunity is meaningful to their business, they will decide to either develop their own internal project to enter the marketplace themselves or acquire the company.

The decision on which path to take will be primarily based upon return on investment; the available IP positions, the product life cycle, and opportunity cost decisions. The return on investment decision is self-explanatory and IP position analysis will be discussed in Chapter 16. The product life cycle management refers to an understanding of where the product is in its market stage. Let's discuss this further.

DOI: 10.1201/9780367533052-5

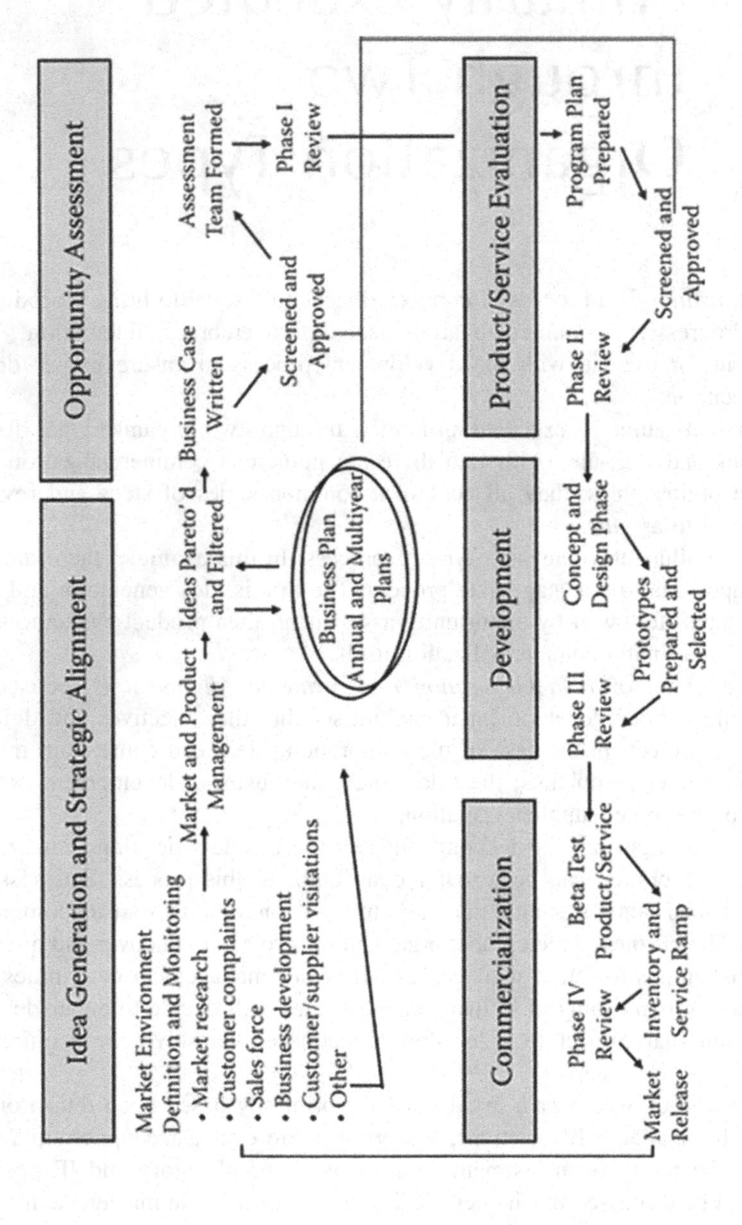

FIGURE 5.1 Stage gate process.

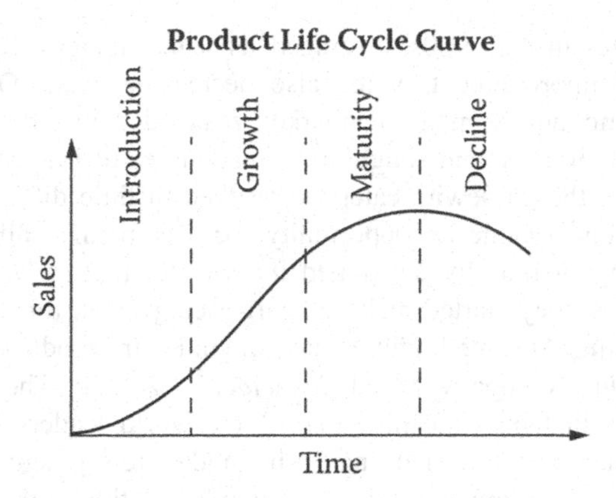

FIGURE 5.2 Product life cycle management.

Looking at Figure 5.2, the *introduction phase* of the market is when something new is being created. There is low competition at this point and firms mostly have losses, as they are investing in developing the market. In the *growth phase* the product starts to grow, which results in rapid sales and profit increases. In the *maturity phase*, sales continue to grow and market share among competitors subsequently starts to stabilize. Market consolidation begins as a major market share is required to maintain profits. Slowing or deteriorating growth rates causes the *decline phase*. Few players remain due to declining profits.

After product life cycle considerations, there is an *opportunity cost* consideration. Opportunity cost is the loss that could occur as taking on the first project may inhibit taking on another. For large corporations, this is not necessarily about having enough money to do both projects; it is frequently about having enough people or internal capacity.

For example, in the medical device industry in the 1980s and early 1990s, corporations such as C.R. Bard, Boston Scientific, Guidant, Medtronic, and others had peripheral (legs) and cardiovascular (heart) interventional products. The primary clinical goals of these products were procedures to get a balloon into the restriction in a vessel and unblock it. The challenge to this procedure was that restenosis (a re-blocking) could occur post-procedure, creating the need for another procedure. During this time frame, the market was in the growth/maturity phase. A new balloon product entering the market that demonstrated an improvement in restenosis rate among their competitors could result in a major shift in market share and profit. There were also other product categories that participated in the procedure such as a guidewire. At this phase of the market, ancillary products such as guide wires were difficult to differentiate and extract price premiums. It was not rational for management to prioritize an internal development effort of a new guidewire over a balloon catheter development project; the opportunity cost was too high given the profits of a balloon catheter and the shift in share a new improved balloon could provide.

The response to this situation was varied depending on the company. For example, C.R. Bard, (purchased by Becton Dickinson in 2017)[1] which had been a leader in guidewires in the 1980s, leveraged its scaled factories and continued to make incremental improvements while also decreasing costs. Other companies, recognizing the imminently maturing market, realized they did not have time to achieve economies of scale and sought external sources of low-cost innovation.

The maturing of the guidewire category resulted in little differentiation among competitors and this became an opportunity for a company called *Lake Region Medical*. Lake Region initially developed its wire technology in making fishing tackle. In the 1960s, they started making heart-pacing wires for Medtronic. In the 1970s and 1980s, they became highly skilled in guidewire production and acquired a leading guide wire company called *Schneider USA, Inc*. The result for Lake Region Medical is that they are now one of the world leaders in guidewire innovation and production.[2] The maturing of the guidewire segment in the peripheral and interventional fields allowed for industry standardization of the category, which was the beginning of a product life cycle for Lake Region Medical. As differentiation of the guidewire category no longer shifted significant market points and profits, business development professions sought a generic supplier as opposed to acquiring a company. As other product categories in the medical device space also move away from differentiation, mergers increase to roll-up these categories to scale. For example, Lake Region Medical's leadership was rewarded in 2015 when Greatbatch acquired the company for $1.73 billion.[3] Greatbatch itself was acquired by Integer in 2016.[4]

The implication of our tangent is to recognize the importance of forming relationships with your exit candidate's business development organizations. The start-up must not only know these individuals but how their particular start-up category is perceived by larger corporations. The start-up must understand the context of business development's view of their product's life cycle. In our example of Lake Region Medical, the maturity of the category of guide wires was not going to lead Boston Scientific, Guidant, or Medtronic to acquire a new guidewire company. Understanding this resulted is validated by reviewing Integer's mission as one of the largest medical device outsourced manufacturers. If a start-up were developing a new guidewire, understanding the market would lead them to see a company such as Boston Scientific as their customer, but Integer as their potential acquirer. More on this discussion later.

Returning to our stage-gate discussion, the next category is *Opportunity Assessment*. This phase requires that the business case be written for the project. In the case of the start-up, this is called the *business plan*—the details of the business plan will be discussed in Chapter 18 later.

The *Product/Service Evaluation* gate is about developing a detailed program plan for all aspects of product development. It is important in developing this plan that you incorporate critical path management (CPM) techniques. CPM is a method of identifying all the interconnected tasks in a project. To maximize the chances of completing the project on time, management focuses on the critical path items, as delays in these items translate to a delay in the whole project. This is particularly important in a start-up, as delays may require raising more capital. *Burn rate* refers to the amount of cash an organization spends per month. The months of delay in a project multiplied by burn rate indicates the additional cash the project could need.

The *development phase* in life sciences can be lengthy, particularly in the pharmaceutical and medical device industries. This is where the product is developed, prototypes are prepared, and clinical feedback is received. Returning to Chapter 1, Figure 1.2, one can see the phases and length of time associated with a drug development effort.

The *commercialization phase* is where the product is beta tested. Beta testing is more typical in the health care IT, biotechnology tools, and diagnostic segments. In the medical device and pharmaceutical segments, completion of the last phase of the clinical trial results in Food and Drug Administration (FDA) approval, which is a surrogate for a successful beta test. During this phase, the company's investment in inventory, marketing, and sales activities expand. All this accumulates to market release.

Although the internal development processes and philosophies associated with commercialization are similar between corporations and start-ups, start-ups have the additional concern of simultaneously raising capital and plotting a liquidity event for their investors.

DISCUSSION QUESTIONS:

1. What is commercialization?
2. Explain the product life cycle curve.
3. Define the opportunity assessment gate process.
4. Define the product/service gate process.
5. What is burn rate and what does it indicate?
6. Reflecting on the Lake Region discussion, why is it important to define the customer and the acquirer? Can it sometimes be both?

NOTES

1 Becton Dickinson Acquisition of C.R. Bard—Frequently Asked Questions for Shareholders, https://investors.bd.com/becton-dickinson-acquisition-cr-bard-frequently-asked-questions-shareholders#:~:text=The%20acquisition%20of%20Bard%20by,wholly%20owned%20subsidiary%20of%20BD.
2 Company Milestones, *Lake Region Medical.*
3 Greatbatch Signs Definitive Agreement to Acquire Lake Region Medical for $1.73 Billion, Globewire, https://www.globenewswire.com/news-release/2015/08/27/763903/10147383/en/Greatbatch-Signs-Definitive-Agreement-to-Acquire-Lake-Region-Medical-for-1-73-Billion.html#:~:text=(NYSE%3AGB)%20today%20announced,%2C%20neuromodulation%2C%20vascular%2C%20orthopaedics%20and
4 Our History, Integer, https://www.integer.net/company/our-history/default.aspx

Section II

Investment Must Be Connected to Exit

6 Angels and Venture Capitalists Invest in Commercialization

THE STAGES OF START-UP FINANCING

Let's pause and take a waypoint—a waypoint is a stopping place on a journey. We will use Figure 1.1 in Chapter 1 as our map, as it illustrates the translation process that investment must go through to deliver innovation. The translation process has macro and micro processes and outputs within it. The previous chapters defined the major macro processes of research, development, and commercialization. The outputs of these processes are invention, patents, and trade secrets, which result in technology. Examining the procedure of cardiac stenting illustrated how it took at least three major technologies to create the product. This demonstrated the cycle of R&D leading into technology and the realization that technology is an input into the New Product Development and commercialization processes.

However, it all begins with investment, as the translation process cannot occur without it. Chapter 3 utilized gross domestic product (GDP) as a global yardstick of economic value. The National Science Board's SEI data on R&D allowed us to appreciate how much the United States invests in R&D as compared to other nations. R&D funding was detailed into its components of basic research, applied research, and development dollars. Each of those components was further broken down into the source of the funding, such as government, industry, and academia.

Next came our first discussion associated with the investment journey specifically associated with start-ups. The Small Business Innovation Research (SBIR) and Small Business Technology Transfer (STTR) programs are specifically designed to assist small companies to conduct their R&D.

We will now complete our investment discussion by exploring how start-ups get funded and where they go to achieve their return to investors—this is called the *liquidity event*. Figure 6.1 provides an overview of the typical investment journey for a start-up company.

The top of Figure 6.1 differentiates the various phases of start-up financing: *Seed, Early Stage, Growth Stage, Later Stage*, and *Exit*. It is important to note that you will hear different terms, but these are the most widely used terms. Let's start by defining the various phases:

Seed is the first stage of financing and the amount needed is generally modest. The goal of this stage is to demonstrate the viability of the business. The definition of viability could be a demonstration that the product could or can work, a demonstration that the market exists, or the hiring of management talent. At this point, there is no commercial operation. The monies are focused on the specific fundable

DOI: 10.1201/9780367533052-6

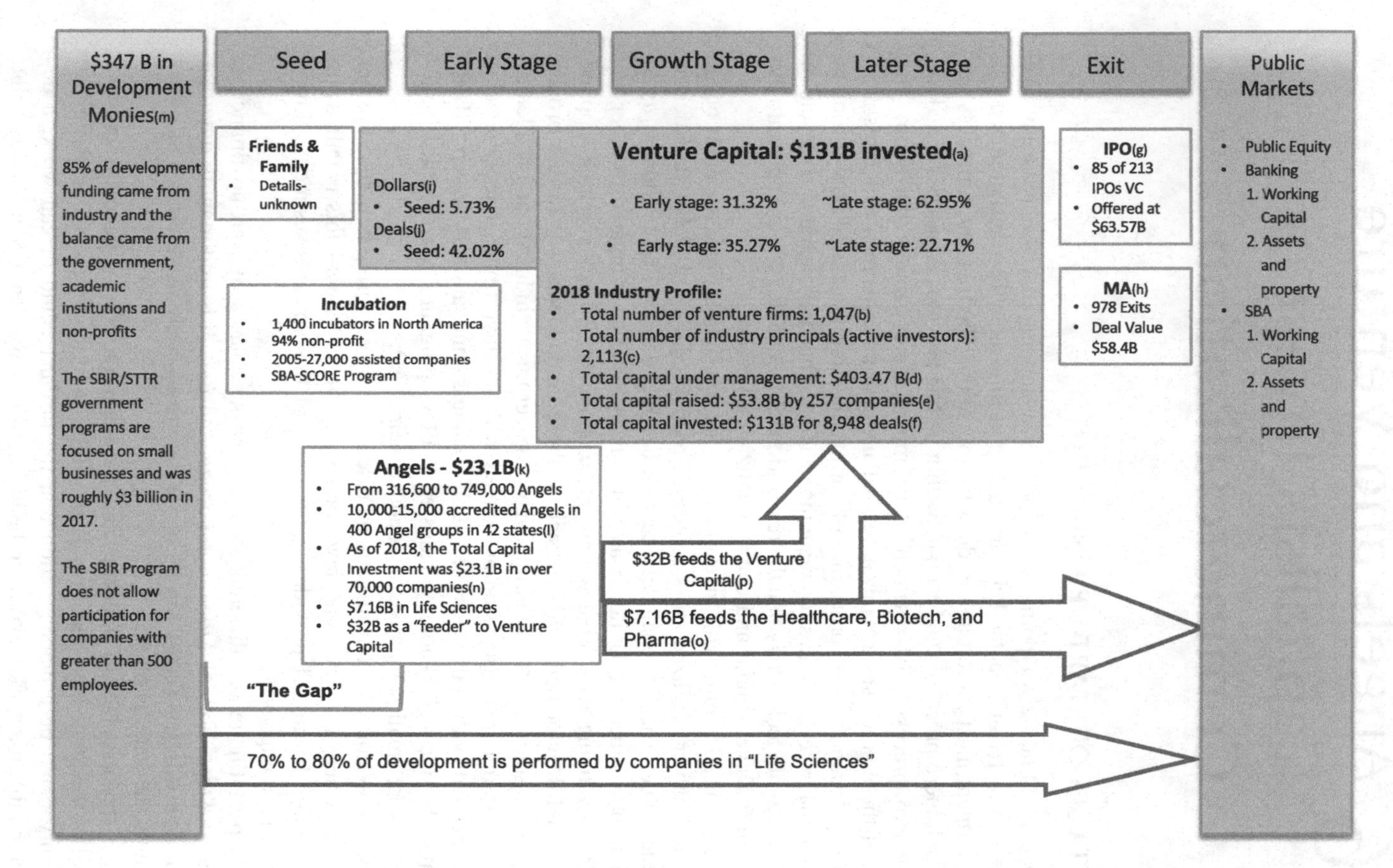

FIGURE 6.1 The start-up capital flow (From James F. Jordan, *Sustaining Life—The Role of Small Business Innovation Research Program*, White Paper, updated 2020, https://healthcaredata.center/technology-creation/startup-capital-flow/. *Source*: Derived data from 2019 NVCA Yearbook, UNF Angel Data, and NSF Data.

milestones that the company must demonstrate to obtain more funding such as demonstrating a proof of concept, validating the market size and competitors, filing a patent, and so on.

Early Stage is when companies are ready to begin operations but are not yet ready to generate sales. In many life science start-up product categories, such as the pharmaceutical and medical device segments, this phase can be lengthy due to regulatory requirements for clinical evidence demonstrating safety and efficacy prior to market approval. With this funding, the company completes their clinical trial, builds out their key management, and finalizes their manufacturing processes. This phase can take 2 to 5 years, depending upon the product category.

The *Growth-Stage* transition from *Early Stage* can be a little fuzzy, as capital required to scale commercial manufacturing could come from the last early-stage funding or the first growth-stage funding. The reason for this is that the company's pilot manufacturing capacity may be sufficient to cover early commercial sales. The transition point generally starts with passing regulatory requirements, or having clinical or economic evidence to support the commercial sale of the product. This phase also includes scaling the sales force and building commercial capabilities.

Later-Stage capital is provided after commercialization and sales but before an exit such as a merger, acquisition, or initial public offering (IPO). The company may be successfully increasing revenues but has yet to achieve a cash-flow positive state—meaning that income exceeds all expenses. The company could also be beyond cash-flow positive, however, the opportunity for an even greater return could be had with additional investment. For example, the return opportunity for expansion into another region or the creation of a new product may exceed the existing positive cash flow and justify raising additional capital for even a greater return than waiting for existing cash flow to fund the expansion. As there can be more than one fund-raising in the later-stage phase, some investment professionals may also refer to these rounds as the *expansion stage*. Others differentiate the anticipated last stage of funding before an exit as a *mezzanine round*.

Exit is the last stage of start-up funding, however, in Section II of this book, it will be emphasized that the exit should be one of the first thoughts during company formation. An exit is about creating a liquidity event for stockholders: the goal being to find a mechanism for stockholders to turn their investment into cash. This can be achieved with an IPO, an acquisition of the company, or a merger. An exit is also about creating a return on the investment for stockholders. Let's pause again and take another waypoint. The reason for the waypoint is that after this discussion, Figure 6.1 reviews the types of investors. The reader needs to understand how to calculate the differing return formulas and, as importantly, know which investor type to use them with. The three predominant methods are *return on investment* (ROI), *return on multiples*, or *internal rate of return* (IRR).

ROI is one of the simplest expressions and is calculated as ([total return less cost of investment]/cost of investment). *Multiples of return*, or simply, *multiples*, is the cumulative returns/investment cost. For example, if an investor received $15,000 for a $5,000 investment, they would calculate this as a 3× return ($15,000/$5,000). The formulas for both ROI and multiples can be criticized for their lack of recognition for time to the return. Time is a critical component in recognizing a return. For example, investing $1 and getting $2 back in 2 years is entirely different than getting $2 back in 5 years. So: why would the simple formulas of ROI or multiples be used?

Venture capital invests into high-potential, high-risk companies that do not have access to public markets or bank loans. A venture capital (VC) fund starts with its *investment charter*. An investment charter communicates to those who invest in the fund the segments the fund will invest in, the roles, responsibilities, and authorities of the various individuals involved, how compensation occurs, how returns will be disbursed, and the expected time frames of those disbursements. A typical VC fund's investment charter is for 10 years (with clauses and penalties for extensions). Venture capitalists target to invest the majority of their funds in the first 5 years and provide returns back to investors in years 5 through 10. Let's return back to our question: why would the simple formulas of ROI or multiples be used? As a VC fund typically expects its returns within a standard time frame it allows for an easy comparison between various investments. With time being fixed, its accommodation is not a material factor for performance comparison between investments or funds.

For nonventure capital investors, time is a material factor, as individual investment returns may occur within different time frames. For these investors, the formula of IRR is preferred as time is considered and allows for comparison between different types of investments. The IRR is calculated by using the net present value (NPV) formula and guessing the interest rate that makes NPV equal zero. Revisiting our previous example above of investing $1 and receiving $2 back in 2 years versus 5 years looks very different through the eyes of time. Take a moment and review Figure 6.2, and observe the differences in expressing a return via multiples and internal rate of return (IRR). IRR is referred to as the *Interest Rate* in Figure 6.2.

Notice that the IRR for 2 years equals a 100% return versus an IRR for 5 years equaling a 14.75% return. These are very different: yet, in the VC world, ROI or return multiple could suffice for comparison, as the time factor for each venture fund is somewhat standard. It is important to note that some VC funds use all three measures.

THE PLAYERS IN START-UP FINANCING

With an understanding of the stages of start-up financing, let's now deepen our discussion on Figure 6.1 to identify the players within each stage, what their motivations are, the risks they assume, and how they make a return.

Seed is the first stage of financing recognized by the private equity community. In Figure 6.1, beneath the heading *Seed*, are three groups of players: friends and family, crowdfunding, and incubators and angels. We need to deviate for a moment to recognize that in many cases in life sciences there may be a pre-seed stage phase. The reason for discussing this topic here, as opposed to including it in our discussion on stages of start-up financing, is that private equity tends not to participate in pre-seed funding; however, some of the seed players may also be involved in some pre-seed activities.

	Return on Investment
Total Return	2
Less: Investment Cost	1
Net Profit	1
Net Profit/Investment Cost	100%

	Return Multiple	Return Multiple
Cumulative Returns	2	15,000
Divide by Investment Cost	1	5,000
Multiple Express in "x"	2	3

Interest Rate	100.00%	14.75%
Initial Investment	−1	−1
Return: Yr 1		0
Return: Yr 2	2	0
Return: Yr 3		0
Return: Yr 4		0
Return: Yr 5		2
NPV	$0.00	$0.00

FIGURE 6.2 Multiples versus IRR.

Pre-seed activities are those required to achieve the fundable milestone for entry into seed stage. In our Chapter 1 discussion of a company named Medrobotics, Inc., we noted that the company was formed out of a Carnegie Mellon University research program focused on snake robotics. Unless there was a personal relationship with an angel or venture capitalist, these investors would not invest in the vision of turning snake robotic technology into a flexible surgical robotic instrument. It is not that they would not like the idea; it is simply that there is too much risk to warrant an investment. Instead, they would monitor the concept and wait for the company to move into a more advanced funding stage, thus de-risking the investment.

So where does pre-seed money for creating that proof come from? Let's return to Figure 1.1 in Chapter 1 and recall that it takes many *applied research* chapters or the process of *translational research* to bridge research to development. A major funding mechanism for pre-seed activities is the National Institutes of Health (NIH) and the NIH's National Center for Advancing Translational Sciences. Figure 6.3 demonstrates the variety of the programs available to fund research and validate individual technologies. The details of each program are beyond the focus of this book and can be found at the NIH's Grant and Funding Web site.[1]

Research Grants

- R01 NIH Research Grant Program
- R03 NIH Small Grant Program
- R13 NIH Support for Conferences & Scientific Meetings (R13 & U13)
- R15 Academic Research Enhancment Award
- R21 Exploratory/Develipment Research Grant Award
- R34 Clinical Trial Planning Grant
- R41/R42 Small Business Technology Transfer
- R43/R44 Small Business Innovative Research
- R56 High Priority, Short-term Project Awards
- U01 Research Project Cooperative Agreement
- K99/R00 Pathway to Indepdendence Award

Resource Grants

- R24 Resource-Related Research Projects
- R25 Educational Projects
- X01 Resource Access Program

Program Project/Center Grants (P series)

- P01 Research Program Project Grant
- P20 Exploratory Grants
- P30 Center Core Grants
- P50 Specialized Center

Trans-NIH Programs

- BISTI - Biomedical Information Science & Technology Initiative
- Blueprint - Neuroscience Research
- Diversity Supplements - Existing NIH Grants & Cooperative Agreements
- Admin Supplements to Existing NIH Grants and Cooperative Agreements
- ESI - New & Early Stage Investigators Policies
- GWAS - Genome-Wide Association Studies
- NIH Common Fund - Roadmap for Medical Research
- OppNet - Behavioral & Social Science Research Opportunity Network
- PECASE - Early Career Awards for Scientists & Engineers
- Stem Cells - Stem Cell Information
- CounterACT Program - Countermeasures Against Chemical Threats

FIGURE 6.3 Government funding mechanisms.

To understand the differences between pre-seed and seed investing, one should think about pre-seed as being about the individual technology and seed being about the product—the pulling together of multiple technologies. The start-up entrepreneur that understands this difference is the one that maximizes their non-dilutive grants. They do so by breaking down the product into as many individual technologies as can receive grants. As importantly, a grant requires formal scoring via a *peer review process.* A peer review process consists of a committee of scientific thought leaders who are expert assessors and score the proposal for its potential impact for the grant authority. Next, the grant authority ranks all of the projects during its assessment period and funds the projects in order of ranking to the extent of their budget.[2] Start-up companies that source their technology and collaborate with academia have the highest potential for non-dilutive funding. However, even those that do not source their technology from academia can use the SBIR program. Utilizing pre-seed funding for independent technology validation not only yields the benefit for entry into the seed stage, it can also be used as a validation point until a patent is issued. Imagine a seed stage company requesting funding from an investor. It is natural for an investor to be concerned that the technology approach may not work or that it is not important or protectable. Until you can afford a patent opinion or actually receive a patent, a grant can be a powerful independent validation source for an investor.

Friends and Family (F&F) refers to investors that are associated with the founders or the management of the start-up company. They are frequently the first investors and an investment return may not be their primary motivator. These investors may be more interested in helping their F&F members get the company started, and as a result, will participate in both pre-seed and seed funding. F&F can also be excited to gain access into this class of stock, as there is not a consolidation point, such as a stockbroker firm, to gain access to seed stage life science companies. For example, in one investment with which the author is associated, the owner of multiple car washes was excited to invest into a life science start-up as he did not know where to find them and was comforted by the fact that his friend knew the industry.

Other motivations for an F&F investor could simply be tax purposes, such as the case of a parent of the start-up founder who was primarily interested in giving his child a start-up experience. Although he hoped for a return, he was equally happy to give his child “a shot” and if it did not work out, he would be pleased to have the tax write-off.

What does gaining access to this class of stock mean? Life science start-up companies require an investment of tens of millions of dollars. Start-up management desires the fewest stockholders with the highest ability to invest to simplify investor relations. As a result of a life science start-up’s high capital needs, lack of a consolidated market, and desire for the fewest stockholders possible, it is difficult for the F&F-type investor to find and/or be invited to invest in a life sciences start-up unless one has access through a relationship. Getting access to a life science start-up company also has the added benefit of the F&F investor having the right to continue their pro-rata investment for the remainder of the project. Pro-rata literally means proportional, meaning that the stockholder has the right to buy their share of future rounds. For example, if an investor bought 5% of the company during the first round, they would have the right to buy 5% of the next round. F&F investors are frequently excited by this prospect. Although the concept of pro-rata share is generally embraced, it is not always the case. In some cases, if you do not always take your pro-rata in a subsequent round of financing, you lose the right to participate in future rounds. In more draconian settings, if you do not participate in your pro-rata share you could lose your preferred stock and be converted into common stock. In other cases, a large investor may ask for the entire subsequent round and demand the cessation of pro-rata; if the current stockholders are unable to continue to fund themselves, the company will compromise its pro-rata policy to gain access to the funding.

Crowdfunding or *sourcing* is another funding mechanism that raises many small amounts via the Internet. This technique has historically been widely used for activities such as disaster relief, nonprofit efforts, and political campaigns. On April 5, 2012, President Obama signed the JOBS Act into law and a component of that law made it legal for private businesses to offer equity to investors via crowdfunding. Although this category for start-up equity has become material in some market segments, it has not yet proved to be a significant funding mechanism for life science start-ups.[3] As the space is evolving there are many Web services that offer crowdfunding services. Since this book's first publication, crowdfunding has grown and, in some cases, different platforms have come and gone. According to an Investopedia article by Mary Kearl in February 26, 2021, Indiegogo, SeedInvest Technology, Mightycause, StartEngine, GoFundMe, and Patreon are the top six sites, the author would add Kickstarted as many of his companies have had success with them.[4]

Incubators are a critical mechanism assisting start-up companies and they can participate in both pre-seed and seed funding. Business incubators are designed to offer programs, resources and, frequently, capital, to help companies successfully develop their products and obtain capital. The International National Business Incubator Association (INBIA) is a good source for specific detailed information and can be found at www.nbia.org. Additionally, your local incubators most likely are also well connected to local angel investors. According to the INBIA, there are roughly 1,400 business incubators in North America and in 2011 they assisted approximately 49,000 start-up companies.[5],[6] The details for incubator types, sponsors, and results are noted in Figure 6.4.

As an example, the author was associated with an incubator called the Pittsburgh Life Sciences Greenhouse (PLSG). The PLSG is regionally focused on Western Pennsylvania and specifically on Life Sciences, which includes the biotechnology, pharmaceutical, diagnostics, medical devices, and technology and health care information verticals. Its success is measured on job growth and wealth creation. Job growth and wealth creation are outcome measurements, and the incubator has identified the four strategies (Figure 6.5) to deliver on its promise of job growth and wealth creation.

The incubator provides various domain-experienced personnel that offer services, programs, and investment money to achieve its mission. The graphic in Figure 6.6 shows how this incubator's efforts flow. The incubator sources its

Incubator Types	
94%	Nonprofits
54%	Mix Use
39%	Technology
4%	Service/Specialty
3%	Manufacturing

Incubator Sponsors	
31%	Economic Development
21%	Government
20%	Academic Institutions
4%	For Profit
8%	No Sponsors
8%	Combination
8%	Other

Incubator Results	
1,400	North American Incubators
49,000	Start-Ups
200,000	Workers
$15 B	Revenue
87%	Still in Business 10 Years Postgraduation

FIGURE 6.4 Incubator industry overview. (From the National Business Incubator Association, State of the Business Incubation Industry, Reporting Years 2006 and 2012, http://www.nbia.org/resource_library/review_archive/1012_02a.php.)

	Need	Solution Strategy
Capital:	Insufficient (local) capital at all stages of life sciences company development.	Increase available capital from multiple sources to accommodate all stages of development, including loans and nondilutive grants.
Connectivity:	Difficulty making critical connections to key resources.	Serve as a conduit to capital, contract research organizations, policy makers, and business development opportunities specific to life sciences.
People:	Lack of experienced life sciences talent, including company executives, managers, and entrepreneurs.	Attract and retain top-caliber technical and managerial talent to support innovation and company formation. Create a pool of serial entrepreneurs.
Space:	Growth hampered by insufficient space, including wet lab space and incubator office space.	Meet the demand for world-class laboratory and office facilities to support the region's life sciences industry.

FIGURE 6.5 PLSG strategies. (From Pittsburgh Life Sciences Greenhouse, Pittsburgh, PA.)

innovation in four different ways. Innovation can be sourced from local universities or entrepreneurs. Technology can also come from larger life science companies that possess IP that they are no longer using or IP that they have maintained for defensive purposes. These companies are motivated to turn these nonproductive assets into cash-generating assets. Last, technology can be imported into the region from any of the three sources discussed: university, entrepreneurs, or company IP (Figure 6.6). The motivation for these companies to relocate is to get the support of an incubator in their commercialization efforts that they could not receive in their regions.

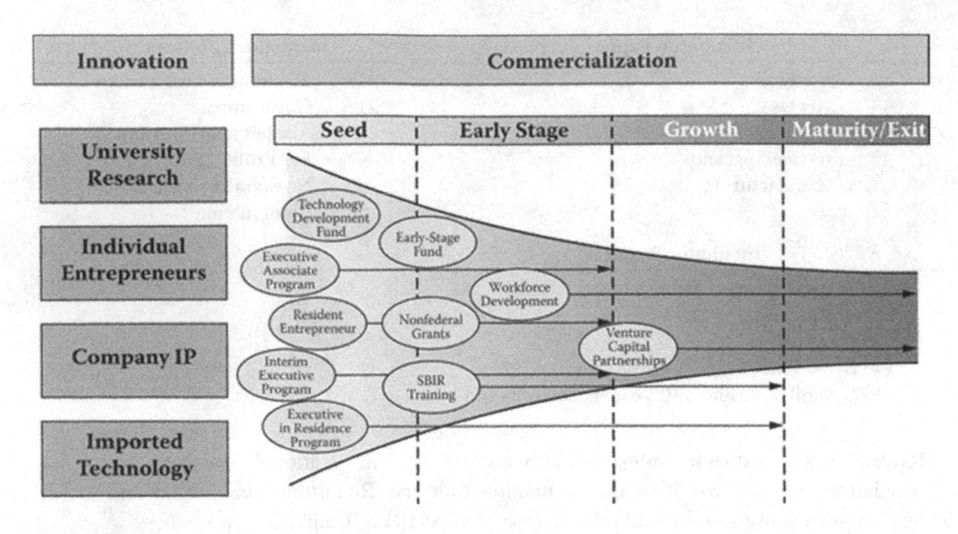

FIGURE 6.6 PLSG investment pipeline programs. (From Pittsburgh Life Sciences Greenhouse, Pittsburgh, PA.)

The incubator's goal is to move companies from the seed stage to exit and, in doing so, deliver on their mission to generate jobs and wealth. A subset of some of the PLSG incubator programs are detailed in the bubbles under the *Commercialization* bar (Figure 6.6). Note that obtaining domain-experienced people and funding the proof-of-concept is the first priority of the incubator. The executive associate, resident entrepreneur, interim executive, and the executive-in-residence program bubbles are focused on getting domain-specific talent into the company to guide its technology development and go-to-market strategy. The reason domain specific talent is so crucial is that, without it, you run the risk of wasting funds on unnecessary activities or extreme inefficiency due to lack of experience. Possible risks include purchasing unnecessary equipment, wrongly focusing marketing and sales efforts, and making poor decisions due to lack of clinical or regulatory experience. As you would not expect your plumber to create and deliver the electrical strategy for your home, company founders are ill advised to allow people without domain experience to plot their strategy. In fact, the franchise industry exists to solve this problem. For example, why would you buy a Dunkin Donuts franchise when you could open a coffee shop of your own? Advantages such as association with a well-established brand, guidance on equipment purchase and facility design, and access to established policies and procedures increase the probability of success. So, to complete our analogy, domain-specific knowledge and relationships are equivalent to franchise knowledge. Continuing our discussion, notice the *Technology Development Fund* bubble: this is about providing funding for the company to invest into developing its technology. This investment usually is in the form of convertible debt—debt that is later turned into stock. Incubators can help set up a company's equity structure and methods to align with industry standards. Note the nonfederal grant and SBIR training bubbles: these are about obtaining non-dilutive funding and an independent assessment of the technology.

The PLSG incubator is just one example of how a nonprofit, industry-specific incubator operates. Other examples include the *IntuitiveX* incubator in Washington and California, which is recognized nationally for their success. Another health care incubator named Blueprint Health in New York is another highly successful health care incubator that provides services and programs but does not provide investment dollars. These are just a couple of examples of how incubator programs and service content may differ according to the resources available in the region. For example, California has numerous experienced life science executives and angels, whereas Pittsburgh does not. Hence, the need to have an executive program and the ability to invest are critical to meet the common goals of all incubators—achieving a successful exit, creating jobs, and generating wealth. To achieve an exit, incubators focus their member companies on achieving a fundable milestone. A fundable milestone is the entry point into the next class of investor. The incubator and the company need to know the specific fundable milestone for the next class of investor whether it be an angel, corporate venture, or VC. Additionally, the incubator must understand that the fundable milestone for each product category differs. For example, in the medical device segment, a proof of mechanical concept may be the entry point for an angel investment and in the pharmaceutical industry a strong *in vitro* (meaning outside of a living organism) study may be adequate.

Earlier we discussed in Chapter 6 the fuzzy seed and early-stage transition. For the F&F and incubators, the goal is to progress the company on to the angel, corporate venture, or VC investors. These investors have the ability to invest the higher capital levels than F&F and incubators are capable of delivering. Returning to Figure 6.1, note the *Incubator, Angel*, and *Venture Capital* boxes. Observe that the *Incubator* box touches the *Venture Capital* box; observe that the *Angel* box extends into the *Venture Capital* box. The reason for the overlap is that much of the time angel capital comes before VC. However, there are occasions where companies can move from F&F and/or incubators' investment into corporate or venture capital. As previously stated, this tends to be specific to each vertical. For example, it is highly unlikely for a pharmaceutical start-up to receive enough F&F and incubator investment to obtain a fundable milestone for entry into the corporate or venture capital class. However, health care IT companies, who generally need less funding, could obtain such a fundable milestone and skip over the angel class into corporate or venture capital.

As *angel capital* is generally the next player, we will continue our discussion with this investment class. Angels use their own personal money to fund a company and in general their individual investments range from $50,000 to $500,000. Many angels have prior professional investment experience and considerable entrepreneurial experience. Angels are individuals who are certainly motivated by a return on investment; however, frequently this is not their only motivation. They may enjoy working with entrepreneurs, may be interested in parttime engagements or may simply want to give back to the community. Angels tend to invest regionally and can be difficult to discover because they do not advertise. Your local incubator and university technology transfer offices are good places to go to get connected to this informal network.

Over the decades there has been an increasing number of angel networks or groups that have formed. These are individuals who pool their money and can invest larger amounts. As angel investors come from a vast array of backgrounds, and a network can consist of many members, there is an increase in the probability that someone from the network has domain expertise. As angel networks have more resources, they generally also have a more disciplined due diligence process, which aids in de-risking their investments. Once an investment is made, the network assigns the best individual or individuals to mentor the company by being board members or advisors. Finding angel networks is easier than finding an individual angel because they advertise their meetings and are members of the various angel associations. The Angel Capital Association has made its member directory public and can be found at: http://www.angelcapitalassociation.org/directory/.

There is another class of angel investor called a *super angel* or *super angels*. Super angels are very high-net-worth individuals who are serial investors. They either operate independently or in a small group. They have a track record of success and are perceived as sophisticated and well connected. Super angels have the ability to invest millions into a company, and if you are fortunate to have one, you most likely will have a higher probability of either exiting without going to venture capital or more easily entering a corporate or venture capital relationship because the super angels have relationships and a track record with downstream investors.

The University of New Hampshire's Center of Venture Research is considered one of the best sources of data on angel capital investment. They have been conducting research on the angel market since 1980. There are between 316,000 and 750,000 individuals that define themselves as angel, however, in any given year since 2002, there have been between 200,000 and 335,000 active individual investors. In 2018, there were 334,565 active individuals.[7] Annual angel investment has fluctuated during this period between $15 billion and $26 billion. In any given year, health care, life sciences, and biotechnology represents between 19% and 30% of angel investment. In 2018, total angel capital investment was $23.1 billion in more 66,110 companies in the United States. Thirty-two percent of angel investment in 2018 was in the health care, life sciences, or biotechnology segments.

The *venture capital* (VC) industry takes financial capital and provides a return to its partners by owning equity in a novel technology or business model that promises high return. Venture capitalists are investors who are skilled at funding and building young companies and they get their money from high-net-worth individuals, insurance companies, foundations/endowments, and both private and public pension funds. Venture fund investors are limited partners and the venture capitalists that run the fund are general partners. The general partners are authorized to run the fund via the *investment charter*. The investment charter usually identifies the industries in which the fund will invest and the stage of investment.

Venture capital generally participates after the seed round: to validate this in 2019 only 5,7% of all venture capital dollars focused on seed stage. Of the remaining 94.3%, 31% of venture capitalist dollars were focused on early-stage investing, 34% were focused on growth-stage investing, and 30% were focused on later-stage investing.

The exact number of venture capital firms can be difficult to determine as funds generally have a 10-plus-year investment charter. Funds typically try to make their initial investments in the first 5 years, and if there are subsequent investments, they are generally follow-on investments into the same companies, rarely adding new ones. In years 5 to 10 of the fund's life, the general partner is typically looking for exits to start returning capital to his investors. When funds are waiting for returns and not investing, they are called *inactive*. However, the fund does legally still exist. So how do you define the number of venture firms active or legally existing? According to a 2019 national venture capital association yearbook, if you captured all firms raising money in the last 8 years, the count in 2018 would be 1,047 firms.[8], Most people consider active funds, funds that are currently making new investments, as being the better yardstick. This makes sense because if you are looking for a VC investor, funds that are no longer investing, referred to as *closed*, are of no interest to the start-up company seeking funding. The last question that could be asked is how many life science and health care funds are active. This is a challenging number to find in publicly available databases as some firms have multiple segment investment charters; for example, IT and medical devices. In our example, the firm could be listed in only one category, such as IT, or in both. Most analysts use the percentage of invested dollars into life science as a surrogate. The purpose of Figure 6.1 is to provide the reader with an understanding of how capital moves through the entire startup ecosystem. To complete this task, a base year for all the data needs to be standardized. However, the 2020 NVCA report has been published and you are more likely interested in the most recent life sciences data. In 2019, 17% of all venture capital was invested into life sciences - down from previous years. However, when looking at the allocation of dollars invested in new deals (meaning first time investments), 31% of the dollars were invested in life sciences.[9]

The last category to discuss is *corporate venture capital* (CVC). CVCs participate in funding rounds in a very similar way to traditional venture capital firms. Unlike venture capital firms whose primary motive is profit, CVCs are interested in investing in start-ups to explore new innovations in their markets. CVCs are particularly interested in innovation that is not part of the company's existing core business and could represent a disruptive change to the company's business model if successful. Like VCs, CVCs are looking at technologies that offer high revenue growth and a return on investment. Figure 6.7 demonstrates the value of the process to corporate venture capitalists and the companies that fund them.

The two charts represent a start-up that needed $40 million to meet its commercialization objectives. The company utilized friends and family investors, angels, venture capital, and corporate venture capital to raise the capital to meet its needs. In both scenarios the company was sold at a 3× multiple to invested equity. On the top example, the corporate venture capital investor experienced the innovation and subsequently the CVC fund's sponsoring company decided not to acquire the company. Upon the company's exit, the CVC got a 3× multiple on their $15 million invested, receiving $45 million in proceeds from the sale, which resulted in a $30 million profit on the deal. In the second scenario, the CVC availed the technology to its sponsoring company, who decided to acquire the company. From the perspective of the CVC's sponsoring company, they bought the start-up

	Company Sold		
Class of Investor	Invested Equity	Acquisition price - assume 3x	Profit by Investor Type
F&F Investors	2	6	4
Angel Investors	10	30	20
VC Investors	15	45	30
CVC Investor	15	45	30
Total Equity	42	126	84

	Company Bought by CVC		
Class of Investor	Invested Equity	Acquisition price - assume 3x	Profit by Investor Type
F&F Investors	2	6	4
Angel Investors	10	30	20
VC Investors	15	45	30
CVC Investor	15	15	0
Total Equity	42	96	54
CVC Savings		-30	

FIGURE 6.7 CVC value.

for $90 million, not the $120 million they would have paid if they had not invested in the company. The sponsoring company gets their $15 million back and spends $30 million less for the acquisition. The previous description is conceptual because in the real world, financial transactions occur within the context of price-earnings (PE) ratios and tax implications.

Another value to the sponsor of the corporate venture fund is the ability to gain privileged information by being an investor or board member. Acquiring a company by being an outsider generally is based upon the start-up attaining certain value or fundable milestones such as demonstration of revenue. For the start-up company, the difference between being bought at Food and Drug Administration (FDA) approval or after demonstration of revenue could be another $10–$20 million in equity. Let's look at the scenario of a company needing $40 million to attain a cash-flow positive status again. However, this time, let's not look at it by investor type, such as angel, but by preferred stock series.

The top section of Figure 6.8 demonstrates the need to invest the total $40 million to achieve the milestones of cash-flow positive and specific revenue achievement in order to be acquired. Investors having the same expectation of a 3× return would want to sell the company to an acquirer for $120 million. In the second scenario, the CVC avails the company to its sponsoring company. The company determines that they could easily put the product through their own sales force and that the project would be sufficiently de-risked at FDA approval. In this scenario, there is no need for the Series D of $20 million. Sticking with a 3× return, the

Class of Investor	Company Sold		
	Invested Equity	Acquisition price - assume 3x	Profit by Investor Type
Series A	2	6	4
Series B	5	15	10
Series C	15	45	30
Series D	20	60	40
Total Equity	42	126	84

Class of Investor	Company Bought by CVC		
	Invested Equity	Acquisition price - assume 3x	Profit by Investor Type
Series A	2	6	4
Series B	5	15	10
Series C	15	45	30
Series D	0	0	0
Total Equity	22	66	44
CVC Savings		-60	

FIGURE 6.8 CVC savings calculator.

company could then be acquired for $60 million versus $120 million, saving the sponsoring company $60 million on the transaction. As the company needs another $20 million to achieve its goals, Figure 6.8 speaks to the savings for the CVC by acquiring the company early. Investors have a 3× return expectation and buying the company earlier saves $60 million ($20 million × 3) in acquisition costs. Referring back to Figure 6.7, additional savings are had for the CVC as they will also not have to pay 3× for the money they already placed into the company.

So, how much does corporate venture invest in life sciences? According to the NVCA, in 2019, CVCs invested into 1776 deals or 24% of all venture capital deals.[10] Since 2014 CVC's have invested alone side venture capital in 21% to 25% of all deals. There is no one organization that aggregates and publishes this data on an annual basis. However, your start-up's business plan may be the best place to look for CVC. The business plan should identify the potential companies that could acquire the start-up. The start-up entrepreneur should look into those companies to determine if they have a CVC arm: it is most likely they do. Why is venture capital delighted to have CVC invest alongside them? The answer is that frequently, CVC's would either be interested in acquiring the company, or interested in the space due to strategic reasons. Thus, CVC's come with great knowledge of the market, and their participation is a signal of a de-risked investment for the venture capitalist.[11]

DISCUSSION QUESTIONS:

1. What are the components of R&D funding?
2. Provide one goal of the Seed stage of financing.
3. What could be the cause for a lengthy Early Stage?
4. Define Later-Stage and give example.
5. How is an Exit or last stage of start-up funding achieved?
6. How is an ROI calculated?
7. Differentiate between venture capital between non-venture capital investors.
8. Define pre-seed investment and how does it differ from seed investment.
9. How is a peer review process done?
10. What are crowdfunding and incubators?

NOTES

1 Office of Extramural Research, Grants & Funding: Research Grants, *National Institutes of Health*, http://grants.nih.gov/grants/funding/funding_program.htm#RSeries (accessed March 15, 2014).

2 NIH Peer Review: Grants and Cooperative Agreements, 2013 Office of Extramural Research Report, *National Institutes of Health*, http://grants.nih.gov/grants/peerreview22713webv2.pdf (accessed June 28, 2014).

3 Kate Taylor, 6 Top Crowdfunding Websites: Which One Is Right for Your Project? *Forbes* Blog, August 6, 2013, http://www.forbes.com/sites/katetaylor/2013/08/06/6-top-crowdfunding-websites-which-one-is-right-for-your-project/2/.

4 Top Crowdfunding Platforms, Investopedia, Mary Kearl, February 26, 2021, https://www.investopedia.com/best-crowdfunding-platforms-5079933

5 Linda Knopp, 2012 State of the Business Incubation Industry, *National Business Incubator Association*, October/November 2012, http://www.nbia.org/resource_library/review_archive/1012_02a.php.

6 University of New Hampshire, Center of Venture Research, CVR Annual Analysis Reports, Years 2002–2012, http://paulcollege.unh.edu/research/center-venture-research/cvr-analysis-reports (accessed March 15, 2014).

7 FAQs for Angels & Entrepreneurs, Angel Capital Association, http://www.angelcapitalassociation.org/faqs/#How%20many%20angel%20investors%20are%20there%20in%20the%20U.S, February 2021.

8 NVCA 2019 Yearbook Public Data Pack, National Venture Capital Association, data provided by Pitchbook, https://nvca.org/wp-content/uploads/2019/08/NVCA-2019-Yearbook.pdf (accessed June 2020).

9 NVCA 2020 Yearbook Public Data Pack, National Venture Capital Association, data provided by Pitchbook, https://nvca.org/research/nvca-yearbook/ (accessed February 2021)

10 NVCA 2020 Yearbook Public Data Pack, National Venture Capital Association, P 32, data provided by Pitchbook, https://nvca.org/research/nvca-yearbook/ (accessed February 2021).

11 Ibid.

7 Create Liquidity for Your Investors

Let's take another pause or waypoint. We started our journey with the understanding that investment uses a translation process to deliver innovation. Innovation, which manifests as either a product, a service, or both, is generally made up of multiple technologies. These technologies require investment and we detailed how research and development monies flow through our economy to deliver technology. More importantly to the start-up, we learned where to source technology and the various non-dilutive funding mechanisms that fund technology development. These non-dilutive funding mechanisms not only provide capital but also help validate the technology prior to receiving a formal patent. This validation is extremely useful in the pre-seed and seed stages of financing as it provides an independent third-party assessment that the intellectual property is important, unique, and has the probability of protection.

Our previous discussions in Figure 6.1 on sourcing technology, obtaining non-dilutive funding, and using an SBIR/STTR to obtain independent third-party validation of the technology are all designed to set the company up to start to attract capital and for the company to appear to be as de-risked as possible given its stage of development. We discussed in Chapter 6 the various stages of start-up financing: pre-seed, seed, early stage, later stage, and the individual players within the stages; friends and family (F&F), angels, crowdsourcing, venture capitalists, and corporate venture capital.

Now that the components of sourcing technology and obtaining investment capital for commercialization have been discussed, it is time to focus our attention toward the end game—the exit. The goal of an exit is to provide liquidity for the start-up's investors. This is generally achieved either through a merger, an acquisition, or an IPO. In 2019, according to the National Venture Capital Association (NVCA), there were 836 mergers and acquisitions and 82 IPOs associated with the venture capital industry.[1] So how can one be obtained? For a start-up to be poised to achieve an exit, the start-up must be a strategic fit, a business model fit, and a cultural fit with the acquirer.

The first question a start-up entrepreneur should ask himself is, assuming that the product or service concept works, how does the solution fit within the industry? This is actually a bigger question than how the product works against competitors. Is the product doing something that has never been done before? Is the product allowing consolidation of an overall value chain? Figure 7.1 presents a flow chart of the health care industry. Although the system is much more complex than detailed in this flow chart, it is sufficient for our discussions on strategic fit and business models.

Using Figure 7.1, let's look at the technology of point-of-care diagnostic testing. When point-of-care diagnostic testing technology became available, physicians

DOI: 10.1201/9780367533052-7

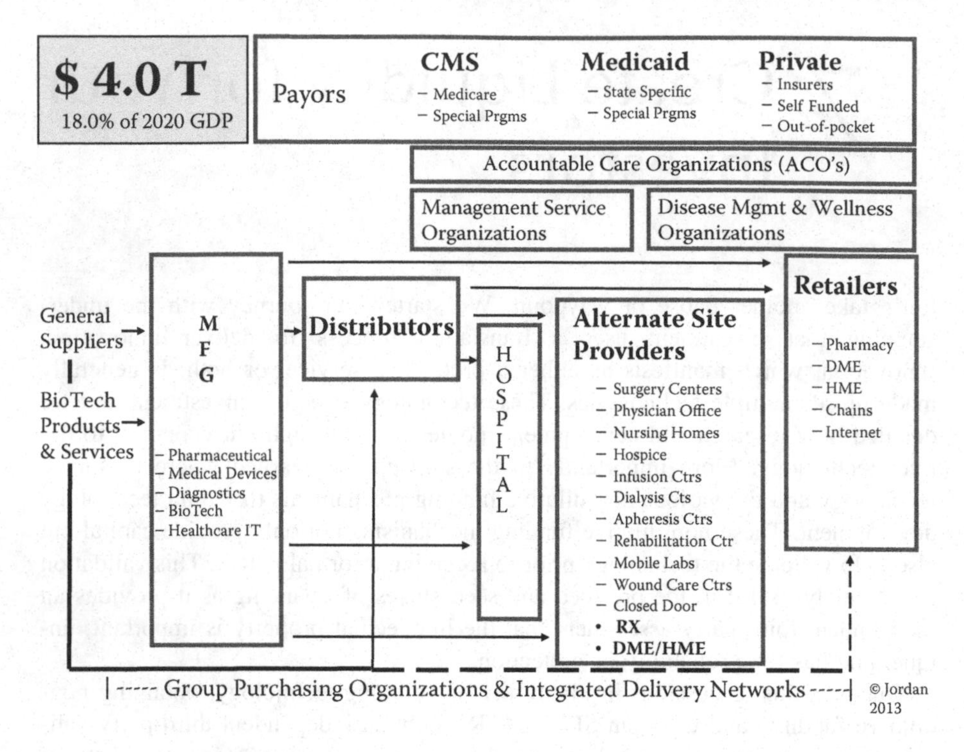

FIGURE 7.1 Health care systems flow chart. (From James F. Jordan 2019.)

were excited to be able to test patients and provide immediate feedback while patients were in their office. An added benefit to the physician was the ability to obtain incremental revenue by being able to charge for the reimbursement code associated with the test. Assume for a moment that your start-up had a point-of-care technology: the story above has clear patient and physician benefit. Why do you think someone would want to buy your start-up? Which companies could buy your company? What value milestone must your start-up reach in evoke action by the targeted acquirers?

Prior to the existence of point-of-care testing (POCT), test samples would either be sent out to an external laboratory for testing or be sent to the central laboratory in the hospital where the sample was taken. The revenue from the reimbursement code would reside with the organization that did the testing.

Two of the largest external lab testers and distributors in the United States are Quest Diagnostics (Quest) and Laboratory Corporation of America Holdings (LabCorp). In 2019, Quest had revenue of $9.4B billion and LabCorp had revenue of $11.5 billion.[2] Let's deviate for a moment and look again at Figure 7.1. There is a box labeled *Manufacturers*, another labeled *Distributors*, and a third labeled *Hospital*. Historically, Quest and LabCorp would buy testing supplies, kits, and equipment from various manufacturers to build their portfolio of testing services and products. They then would receive samples, in this case from hospitals, perform the tests, and distribute the results back to the physicians. In another scenario, these companies could also deliver the testing supplies, kits, and equipment to the hospital central labs. This detail demonstrates that these companies have significant control over the traditional diagnostic sales channel.

Let's consider how these two companies might have initially looked at POCT. Prior to POCT technology, Quest and LabCorp had a major presence in the two locations for traditional testing: external labs and hospital central labs. They dominated the revenue associated with these categories of tests. How could these two companies look at the existence of such technologies?

From a strategic perspective, these POCT technologies could be seen as both the creation of a new category and the rearranging of the value chain. Testing done in the physician office could cause a loss of revenue for both Quest and LabCorp's traditional testing business. Additionally, it could potentially cause a loss of revenue to a major customer—hospital central labs. Thus, Quest and LabCorp would need to decide if this threat would meaningfully detract from their existing core revenue stream. If the result were deemed meaningful, they would need to develop or acquire POCT technologies to complement their traditional lab revenue business. However, they could also risk losing core business from the hospital's central lab if the central labs deemed Quest or LabCorp were promoting POCT and taking business away from them. If they felt it was not a threat, they could ignore the category or attempt to discourage its formation by inhibiting access to their sales channels. They could also view this technology as an anomaly and either ignore it or watchfully wait to determine if the category developed and acquire the capability at a later date.

The following milestones on the HemoCue Web site offer insight into how this start-up evolved over 30 years[3]:

- 1974—The new method for improving the measurement of hemoglobin.
- 1979—A license agreement with pharma company LEO is signed.
- 1982—LEO begin selling and distributing the first hemoglobin for meters.
- 1986—The hemoglobin system is introduced to the U.S. market, revenue rises 80%.
- 1990—A U.S. distributorship is formed.
- 1992—HemoCue revenue $47 million, sold to Mallinckrodt, Inc. for $100 million.
- 1999—HemoCue acquired by a private equity group called EQT.
- 2003—Distribution agreements with McKesson and PSS.
- 2007—HemoCue acquired by Quest for $420 million.
- 2013—HemoCue acquired by Radiometer for $300 million, Quest to refocus on diagnostic information services.

Now that we know HemoCue's history, let's take a look at the status of the POCT market today and see if we can derive some insight. According to Frost & Sullivan, the global revenue in 2012 for the POCT market is estimated to be at $5.3 billion and expected to grow to $9 billion by 2019. This is a compounded annual growth rate of 7.9%. Companies offer their products through both direct and distributor sales channels. Taking a look at the market share leaders in the 2012 Global POCT Market is illustrative[4]:

- Alere (Acquired by Abbott 2017) —39.7%
- Danaher Corporation (Radiometer)—16.0%
- Roche Diagnostics—11.0%
- Siemens Health Care Diagnostics—9.3%
- Instrumentation Laboratory—8.8%
- Others (more than 70 players)—15.2%

According to various industry reports, including Frost & Sullivan, the following environmental elements impact the POCT market:

- If test results are used to alter patient therapy, FDA approval is required.
- Lack of insurance coverage and poor reimbursement rates to point-of-care (POC) physicians.
- Central labs (in hospital) still have the cost advantage.
- Connectivity of diagnostic information to connect to an electronic health record.
- The transition from traditional testing to molecular diagnostics.

Now that we have a sense of how the diagnostic industry flows (Figure 7.1), who the two major traditional diagnostics players are, the history of HemoCue, and the current size and status of the industry, we can, through the benefit of hindsight, review the situation.

HemoCue was developed in Sweden by scientists who possessed domain-testing experience that allowed them to develop a proprietary approach to POC hemoglobin testing. Their scientific knowledge and intimate understanding of their environment resulted in strong intellectual property. It was the company's strong IP that created the unfair advantage that allowed for an exit.

The negative of the situation was that the larger market was in the United States and the scientific founders had no U.S. diagnostic business experience. This lack of experience inhibited the founders from specific knowledge of their business' strategic fit within the industry and the business model's fit with its potential acquirers. As you can see from history, the company was aware of this gap and they subsequently hired selling and distribution organizations.

In 1992, HemoCue, with revenue of $47 million, sold the company to Mallinckrodt for $100 million. At this point, the founders of the company certainly created an exit. The next question is, was it a profitable exit, and if it was profitable, was the opportunity maximized? Information on the exact amount of return to investors is not public, however, it would appear on the surface given a 2× multiple to revenue that the company most likely made a positive return on investment to its stockholders. However, was the opportunity maximized?

Given that 15 years later, in 2007, the company was sold for $420 million to Quest, it is highly probable that the opportunity was not maximized. Quest's willingness to take a $120 million loss on the sale of the company to Radiometer provides further evidence on the lack of strategic fit within the traditional diagnostic industry. Looking at the 2012 POCT market participants, Radiometer, HemoCue's current owner, has a 16% share of the POCT market. It appears that HemoCue has now been acquired by a company who not only understands the POCT market but has a meaningful market position within it at 16%.

Why didn't Quest or LabCorp buy the company in the 1990s? Did they fear alienating their central lab customers? Did they anticipate a difficult reimbursement and regulatory environment? Did they anticipate that this category had a limited market size? They most likely knew the future of molecular diagnostics was anticipated and that these tests are difficult to perform in a POCT environment.

From the perspective of a strategic fit, business model, and culture, there was not a fit. From a business model perspective, Quest and LabCorp were most likely not going to risk their core business (lab testing services) and their core customer (central labs) for an emerging category. Their business model is the accumulation of thousands of individual products and services that make up their multibillion dollar revenue streams. Returning back to Figure 7.1, we can understand the historic culture as manufacturers came to Quest and LabCorp with products and in turn, if the products demonstrated revenue traction, they would be put through the channel. An acquisition was not generally needed given the company's sales channel control, as, if you wanted to get your products to their customers you were required to go through them. So why did Quest ultimately buy the company?

The 2003 HemoCue distribution agreement with McKesson and PSS (now owned by McKesson) may provide some insight. HemoCue's success with McKesson and PSS likely caused concern to Quest that POCT could go through other types of distributors. Additionally, today's market leaders, as identified in the 2012 market share leader charts, both sell their equipment directly to the customer and/or using other types of distributors. This may have caused concern to Quest that a major category was capable of being controlled by another distributor sales channel. However, it appears that Quest never asked itself the questions of strategic fit and business model. Although Quest may distribute some equipment, has it ever been in the business of manufacturing equipment?

Another question: Why did HemoCue sell to Mallinckrodt, as Mallinckrodt did not have a significant presence in this market? One possible answer was the company was 18 years old and it was likely that the investors were anxious for liquidity. Another could be that Mallinckrodt was reforming itself at that point in time and this was to be the foundation for a future POCT business. We will never know all the answers.

This discussion is based upon publicly available data. It is purposely void of the academic analysis, confidential data, and private interviews that accompany a Harvard Business School case study. The entrepreneur generally does not have the resources to conduct such a sophisticated analysis. If the data is available and affordable, by all means get it. What a start-up is capable of doing is asking the basic questions of who is going to buy this company and why. Once the list is created, the next filter must be an analysis of strategic fit, business model fit, and cultural fit. The start-up must scrub publicly available data and find domain-specific people to answer these questions. The start-up should not proceed to raising capital without having this discussion.

DISCUSSION QUESTIONS:

1. In Figure 7.1 Health Systems flow chart, who are the three main payors?
2. What is the goal of an exit stage?
3. What is the first question that a start-up investor should ask if a product/service concept works?
4. According to various industry reports including Frost & Sullivan, provide as least 3 elements that impact the POCT?
5. Once the list of basic questions is created by a start-up investor, what should be the next step?

NOTES

1 NVCA 2020 Yearbook, Data provided by Pitchbook, In Review, p. 36.
2 Hoovers Online Database 2013, *Hoovers*.
3 30 Years of Milestones, *HemoCue*.
4 Frost & Sullivan, *Global Point of Care Testing Market, M8FD-52*, May 2013, Competitive Analysis—Market Share, May 2013, pp. 71–72.

8 A Liquidity Event Is Not Consummated without Due Diligence

A START-UP IS DESIGNED TO BE TEMPORARY

A start-up company is designed to be a temporary organization. Its goal is to uncover a business opportunity that can be exploited by an unfair advantage. An *unfair advantage* is a combination of unique personnel talent, proprietary relationships, know-how, assets, or IP that deliver a repeatable and scalable business model that is not easily replicated by others.

Successful start-ups have experienced management that understands how to navigate the complexity of health care. It is also important that the management team have experience in connecting innovation sources with capital sources. Management must be able to obtain technology (frequently from universities, invented themselves, or other companies), and simultaneously have the ability to connect with capital sources (investors).

The final objective of the start-up is to provide wealth for its investors by providing a *liquidity event* that delivers an *unusual return* for its investors. A liquidity event is the conversion of the stock into cash that is achieved by the sale of the corporation or an initial public offering. The term *unusual return* is used to indicate that investors expect a much greater return than what they would receive in the public stock markets. Investors balance risk with reward and a start-up is a much riskier investment than most endeavors. As a result, investors naturally expect a much greater return.

THE REASONS FOR DUE DILIGENCE

Due diligence is a generic term used to communicate the need to research either a business, a group, or an individual prior to forming a formal relationship. Its legal intent is to demonstrate that a level of care or prudence was taken to ensure that you validate that you will be getting what you thought you were in the relationship. Due diligence is generally done prior to signing a contract, investing, or acquiring a company.

As a start-up company by definition does not have a long-term track record of performance, the process of due diligence will be a continuous one. The company should expect numerous constituencies, such as suppliers, investors, and customers to perform a stream of due diligences on the company over its lifetime. This succession of due diligences hopefully leads to the final due diligence—the company's acquisition or initial public offering.

DOI: 10.1201/9780367533052-8

Unfortunately, most start-up companies do not see the continuous nature of this due diligence process. As a result, they see this important activity as an event initiated by a specific due diligence request. Equally alarming, many of the entities that request a due diligence process have an ambiguous process themselves. This unplanned approach brings risk to the company. Not being prepared for due diligence can, at best, delay the completion of the due diligence activity; at worst, it demonstrates that the company cannot back its claim, or unintentionally presents contradictory evidence, causing potentially irreversible harm to the company.

The latter situation can immediately cease the formation of a critical relationship and adversely impact the company's reputation. An event worthy of due diligence is frequently an event generally associated with a value milestone. A value milestone is a waypoint along the journey toward an exit. Not consummating the due diligence can impact the company's chance of achieving that milestone. Additionally, confidentiality agreements seldom protect industry insiders from learning that a particular company performed a due diligence. A failed due diligence that results in a failed deal generally sends a negative message to the industry. This message can impact the company's momentum in forming other relationships necessary to achieve that value milestone, as industry insiders assume there is something wrong with the company. This can result in a major setback.

When one considers all of the above, due diligence is a major business process and an important component of a company's brand: a successful due diligence supports the company's brand; a failed due diligence can damage it. It therefore warrants the same attention to process and quality control that the company gives to manufacturing.

RISKS ASSOCIATED WITH DUE DILIGENCE

A risk is a situation that may cause danger, harm, or loss to the start-up company. In their pursuit of a value or fundable milestone, start-up companies frequently overlook the risks associated with inadequate preparation for due diligence. Ironically, this lack of thoughtfulness may offset the value sought, or more severely, cause a decrease in value. Even if the company is cautious of due diligence risk, the lack of a systemic approach leaves gaps. Gaps frequently result in not considering all constituencies in the risk equation and the individual messages they send to each constituency. In Chapter 9, Figure 9.1 offers a "Due Diligence Checklist." This checklist can also be downloaded from the book's Web site, at: https://healthcaredata.center/commercialization-startups. The objective of Figure 9.1 is to demonstrate the many facets of constituency risk that should be considered. Given the size of the document, an overview graphic is provided in Figure 8.1.

Even if the company adequately prepares to have a due diligence performed on itself from investors, regulators, and even customers, they must be equally prepared to perform due diligences on others such as a supplier or contractor. The start-up company must be cognizant that their due diligences are *discoverable*, meaning that other parties may request the details of the start-up's diligence. Thus, the quality of the company's due diligence on its supplier may become part of the due diligence that is performed on them. The comprehensiveness of the supplier's due diligence

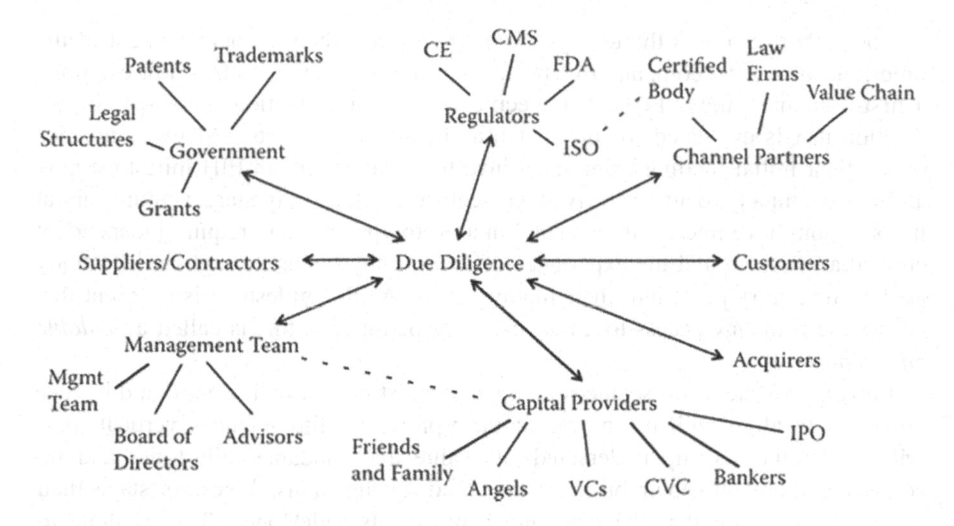

FIGURE 8.1 Due diligence social network.

process could be a telling surrogate for the group performing the due diligence on the start-up, regarding the quality of their systems and the experience level of their management team.

Unlike other start-up categories, life science companies (with perhaps the exception of Health Care Information Technology [HIT]) can take 5 to 12 years and $15 to $150 million of investment before achieving positive cash flow or an exit. As a result, demonstrating achievement of the next milestone unlocks the ability to increase the company's value and obtain access to additional start-up capital.

Frequently, an independent third party, such as the FDA, the U.S. Patent and Trademark Office (USPTO), a specific partnership, or some other branded investor validates achievement of the next milestone. Independent validation of a milestone brings great confidence to the organization that is performing due diligence because its *arms length* assessment is independent from the company. For those not yet moving to due diligence but observing the company, it also sends a message that the company is progressing and has continued to de-risk itself.

The large amounts of money necessary for life science companies to exit requires them to frequently traverse angel capital, venture capital, and venture and banker debt categories of funding. As it is necessary for the life science category to reach across various funding classes, the author differentiates between a value and a fundable milestone.

A milestone marks increments of distance between the beginning and end of a journey. Each life science start-up category—biotechnology, pharmaceutical, diagnostics, medical devices, and HIT—all have milestones that are specific to their category. These milestones could be the building of a prototype, obtaining a patent, achieving *in vitro* or animal testing, and so on. Attaining one of these milestones generally increases the value of the company: hence the term *value milestone*.

The author also uses the term *fundable milestone* when the achievement of the milestone allows the company to advance into the next funding class. For example, a first-in-human study (FIH) is a specific clinical investigation for a specific indication that is evaluated for the first time in human subjects.[1] Many companies obtain their initial funding from angel investors. Achieving an FIH milestone may allow the company to attract early-stage venture capital. Early-stage venture capital firms seldom have interest in investing in a prototype and may require independent clinical assessment and the experience of several physicians through an FIH study as their first entry point into their funding class. As this milestone is an event that allows the company access to a new category of funders, this is called a *fundable milestone*.

Previously stated, life sciences companies must design their own due diligence processes to align with the norms of their particular life sciences vertical. It is critical that the start-up understands its value and fundable milestones and incorporates them into their business and fund-raising plans. Investors stage their financing to ensure that the company achieves its milestones. This is done as achievement of the milestone signals to the equity investor that the management team members are capable planners and that the company is less risky.

If the milestone is not met and the company does not have sufficient funding to continue, it must raise additional funding from existing or new investors. As previously discussed, each industry vertical has a value for each specific milestone. If the company needs to raise more funding to get to the milestone, and the milestone's value is less than the previous company's stated value, investors generally request a decrease in the firm's prior value to accommodate for the misstep. This is frequently called a *down round* because the value of the last equity round is decreased to accommodate for the misstep.

The degree of consequence to existing stockholders is frequently determined by whether you are at a value or a fundable milestone. If you are at a value milestone the company's existing stockholders may be able to fund the company without needing new investors. However, if the failed milestone was a fundable milestone, meaning the company was about to traverse into a new investor class, the consequences can be devastating.

There is, perhaps, no clearer example than a start-up pharmaceutical company where millions of dollars are necessary to achieve the next milestone. Should the fundable milestone be the transition point from an angel investor class to an early-stage venture capital class, the results can be particularly devastating to the angel class investors. If the angel class investors do not possess enough capital to fund the company, the early-stage investors will see this as an opportunity to acquire an inexpensive investment. For example, if an early-stage venture investor model starts at a typical valuation of $17 million and they can now get that same investment for $12 million, they have de-risked their investment. The reason they have de-risked their investment is that if the typical valuation point is $17 million and they got it for $12 million, they have $5 million ($17 million – $12 million) that can accommodate for either future missteps or become an unusual profit for their own investment portfolio in the future.

Management must recognize the dire consequences of failing to incorporate the risk of missing milestones into their due diligence process. A down round not only devalues the existing investors' stock but also the stock reserved for the management team. Missing a milestone at the wrong waypoint, such as a fundable milestone, can stop the company from progressing into a more advanced class of wealthier investors. If this happens and the existing stockholders do not have enough funds available to achieve the milestone and advance to the next class of funders, the results can be devastating. Should the company find a way to survive, the management team may not, as investors may have lost confidence in management's ability to manage its milestones and risk.

Building a solid company reputation is no different than a solid ice surface forming on a lake. When it's cold enough, small molecules solidify. Given time and the right conditions, they crystalize to each other in unpredictable arrangements to form a solid surface. However, if those conditions are not maintained, at best the surface weakens, and at worse, the solid surface turns back into water.

The analogy here is that a company's reputation must be tended to in order to create a solid foundation for an exit. New relationships must be forged and expanded. Reputational harm disrupts and weakens one's investment in those relationships. Start-ups must tend to these relationships with the same diligence as a company would put into its product's design.

A real-life example demonstrates how fragile managing a reputation can be. The J.P. Morgan Healthcare Conference is one of the biggest annual events that consolidate life science executives and investors into one location. This annual event occurs in January and is held in San Francisco, California at the Westin St. Francis. The lobby at the Westin has become a popular location for life science executives and investors to meet prior to having a meeting. Several years back while waiting to meet a colleague in that lobby, I was standing next to a gentleman named Bill. We both exchanged pleasantries while we awaited our guests.

Bill's guest arrived and I could not help but overhear their conversation. The gentleman asked Bill if he had heard about a company that my organization had invested in. Bill stated that he knew this company well. The gentleman then asked a follow-up question, "Will you be investing in them?" I was surprised to hear Bill say that he would not.

Upon hearing this, my curiosity exceeded my politeness, and I asked Bill what he thought was wrong with the company. Bill stated that there was nothing wrong with the company and in fact he liked it. I followed with another question, "Why aren't you interested in investing?" Bill stated that he did not invest in companies while they were in clinical trials, only after regulatory approval. I immediately turned to Bill's guest and asked, what would he have thought of Bill's opinion of the company if I had not asked this additional question. The gentleman said he assumed Bill was negative on the company and he was going to stay away from it. I followed up with another question to Bill's colleague, "If someone asked you your opinion about this company, what would you have said?" He said he simply would state that he was not interested in investing in the company. I asked what would have been his reason why if he were asked. He chuckled and said most people would not have asked the second question. He commented that he gets hundreds of business plans a month and cannot read them all. He needs to narrow them down quickly and he uses his network of colleagues to help him determine what plans he is going to read.

One can appreciate from this story how quickly a false impression can spread. The life sciences industry is a close-knit community of executives, who are the buyers; and investors, who are the sellers. These groups communicate with great frequency and sadly, the communication is frequently as cursory as the one described above. Thus managing impressions is the most important thing a management team can do for their company.

Favorable impressions can spread as quickly as false ones and obtaining milestones and having favorable due diligence outcomes best achieves the acceleration of impressions. External organizations favor milestones and due diligences because they are validated by independent third parties. Obtaining regulatory approval, filing or obtaining a patent, forging a strategic partnership, or signing a major distributor are all signals that can only be sent with the consent of an independent third party. It is the independence of these third parties that publicly validates the company's progress.

As important as it is for the company to be prepared for a successful milestone achievement or due diligence, the company must also ensure that the independent third-party's reputation is also of the highest caliber. The validation of a milestone or due diligence is only as good as the reputation of the independent third party. Start-up companies must be cognizant of the importance of both of these factors in creating reputational momentum. Companies must enter into these processes with the kind of preparation that ensures the highest probability of success.

We have previously discussed that start-ups are judged by a series of waypoints as they journey toward an exit. Each waypoint is valued as a milestone and fundable milestones require particular attention as they provide entrance into new classes of investors.

We have discussed that each vertical has its specific milestones toward exit and each milestone has an expectation as to how much money should have been spent to achieve it. Start-ups must also be cognizant that these milestones are also measured against time. The same information technologies that allow value comparison at each milestone also allow for time comparison. Time may be an indicator of risk. Sir Isaac Newton's laws of motion provide an interesting framework in which to search for risk due to time delays.

Newton's first law of motion is the law of inertia, which states that an object in motion continues in motion with the same speed and in the same direction unless acted upon by an unbalanced force.[2] The inability of a start-up to achieve its milestone within the allotted period of time may suggest that a company has either lost its internal momentum or an unseen external force is acting upon it. As outside investors or acquirers are looking for signs of risk, Newton's first law causes pause to discern if there is an internal problem or an external problem that is inhibiting progress. For example, perhaps internally there is an existing investor with material voting rights who will not allow the company to seek the outside funding for fear of stock dilution. Their fear may be inhibiting the start-up's management team from obtaining the resource to move the enterprise forward. An external example could be that the company requires a partnership to obtain rights to an important patent and the future partner has unrealistic expectations of compensation for that right. These companies both may be at an appropriate milestone but the time it took them to get there causes the due diligence organization to pause to look for unseen risks.

Newton's second law of motion is that acceleration is produced when a force acts upon mass.[3] In short, heavier objects require more force to move than lighter objects, given the same distance. Frequently, this law can manifest itself when a company achieves a milestone within a time frame that is not only longer than standard but also cheaper than standard. This can be an indication that the company does not have sufficient internal resources to achieve the subsequent milestone. Each forward milestone meets increasing market forces: one would expect that the start-up requires increasing resources to continue its momentum. A frequent culprit violating this law occurs in the 510k category of the medical device business. The 510k category seldom requires a significant clinical trial. Attempting to be frugal, the medical device start-up creates its prototype, submits it to the FDA, and attains approval to commercialize its product. The challenge comes in when the start-up feels FDA approval is their end point. The start-up attempts to pursue an exit and soon discovers that if they get an offer to be acquired, the offer is significantly under (30%–60%) what they believe the standard exit would be for their company. The company is now faced with a dilemma: sell itself for under standard value or continue on to its next milestone. In this example, the next milestone is to generate meaningful revenue. The problem is that the company never planned to sell the product, so not only do they not have the personnel to generate that revenue, they do not have the funding to obtain the personnel. The company, unprepared to raise capital, now must do so. Unless the company is in a position to raise capital from its existing investors, they are most likely going to have a down round. External investors are not only motivated to offer a down round to get a bargain but they are equally motivated for a down round to offset the risk of the poor management that was demonstrated in getting into the predicament in the first place.

Although an independent data source is not available to offer statistics, it has been the author's experience after working with 400 start-ups that up to one-third of these companies will fail at this moment. The failure is caused by existing investors not having the funding to continue and more advanced investors seeing the predicament as a management issue, so they fear the upside of a down round does not offset the demonstrated inexperience of the management team (management risk).

There are obviously exceptions to these rules. There are companies that prepare from the very beginning to take it slow. These entrepreneurs decide from day one to self-fund the company by using either their own personal funds and/or from the operating profits of the company. This approach is frequently called *bootstrapping*.[4]

Successful bootstrappers tend to be either niche market players or obtain their strength from their intellectual property (IP). Market timing does not impact their value as much as traditional start-ups because their place within the market is secured because of their niche or IP. As they planned to bootstrap from the very beginning, they avoid the pitfall of running out of cash. Once these companies have demonstrated their value through revenue, they seldom need to accept a low exit offer because the acquiring company knows they are capable of continuing to self-fund their growth.

Figure 8.1 shows that what may be perceived initially as a simple process quickly becomes grand and complex. With such complexity, the risk of being incongruent increases. Incompatible stories avail themselves quickly.

Cloud-based software technologies have improved communication among investors. Available software systems such as AngelSoft and VentureSource, to name a few, allow for the communication of milestones and valuations among investors. The software systems allow for comparison and benchmarking among similar companies. Companies must be aware and prepared to measure themselves against these benchmarks. It is more difficult to argue your value if you are operating against the norms of your vertical.

Information technologies have also improved communication among regulatory bodies as they seek intelligence on applicants. Increasingly this is happening to life science companies, as incongruent communication is being detected and interpreted by individual regulators. Without knowledge or rationale for the contradictory information, start-ups run the risk of stalling, or worse, finding themselves in a situation where their business plan must change to resolve the situation. Changing a business plan midstream frequently is costly, if not deadly, to the company particularly if it interrupts the achievement of a fundable milestone.

No clearer example exists than in the medical device space, where companies seeking FDA section 510k clearance may be sending a contradictory message to the USPTO. A 510k is a pre-market notification process to the FDA that states that the company's device is substantially equivalent to a similar product that was commercially available prior to May 28, 1976.[5] The USPTO defines patentability as something that has not been previously publicly disclosed.[6] You can immediately sense the contradiction between the 510k's substantial equivalence (a statement that it is like something else) with the USPTO's patentability definition (it's not like something else).

The FDA also sees the contradiction and the motives as medical device start-ups frequently prefer the 510k pathway to commercialization as it is significantly shorter and less expensive than its counterpart, the Pre-Market Approval (PMA). The PMA process is the most stringent type of device application in that it must contain scientific evidence to ensure that the specific device is safe and effective for its intended use. Through a PMA, the FDA is thus approving a specific device utilizing its specific data.[7] Again, you can immediately sense that the FDA may expect the filing of a patent to align with the requirement of a PMA application, not a 510k.

When one thinks of a start-up company, one most likely thinks of an equity investment. An equity investment allows one to receive a stock certificate representing a piece of ownership in the company in exchange for their cash. Frequently, the risk and uncertainty of the start-up requires the use of other mechanisms to reduce risk for early investors.

We previously discussed in Chapter 13 the concept of a down round, which is an equity penalty for a failed milestone. Our discussion here is overly simplified to express the basics of a failed milestone on equity ownership. The complexities of a real-world down round will be discussed later.

For now, let's assume the company originally planned to value itself after reaching milestone #1 at $50 million: the Preferred A class stockholders would have $50 million of value. Let's also assume that the company planned on raising another $50 million to get to milestone #2. At milestone #2, the company would be valued at $100 million. Preferred A and Preferred B would each have $50 million in equity value.

However, in this scenario, the company missed its milestone #1 and now needs to raise $60 million, which is made up of $10 million to complete milestone #1 and $50 million to complete milestone #2. Applying the rule of simple fairness, the new Preferred B class would expect to have 60% of the milestone #2 that is anticipated to be valued at $100 million. This is required as Preferred B is investing to both complete milestone #1 and milestone #2. The remaining 40% of the value is left for Preferred Series A.

What this example is missing is the penalty for risk. Our previous calculation would be rare in the real world, as it does not recognize the risk Preferred Series A demonstrated in their mismanagement or poor planning. Our previous calculation did not incorporate this demonstrated risk and must be assumed to continue to be present as the company marches toward milestone #2. Thus, the Preferred Series B investor would be prudent to extract additional value to compensate for this potential risk. This ends up being a negotiation and it would not be unusual in our example here to have the Preferred Series A only owning 20% to 30% of the company at milestone #2.

The problem with the above example is that it is a negation that has no independent, third-party calculation of risk. It is a negotiation between two parties that are not capable of calculating the exact cost of the risk—it is unquantifiable.

Warrants are a good mechanism to resolve the debate. A warrant gives its holder the right to purchase additional equity at a specific price within a certain time frame.[8] When a milestone is not met, warrants may be used in a few ways to compensate for risk. Warrants may go with a particular round of funding to incentivize investors to invest or warrants can be used to compensate for risk and expire when the risk is removed. The removal of the risk is usually expressed as the achievement of a specific milestone, at a specific time, within a specific budget.

Continuing our example above, a company may add warrants to its stock offering to reduce the risk profile for the new investors. Suppose a company has a Preferred A class of stock and a Preferred B class of stock—each owning 100 shares of the company. If the company were to be sold each class would expect to receive 50% of the proceeds of the sale.

However, let's change this example and add 50% warrants to the Preferred B class of stock. If the Preferred B class exercised its warrants of 50% they would take their original 50 shares and multiple them by 50%. They would now have the right to buy another 25 shares (50 shares × 50%) at a predetermined price. The company's stock ownership structure has changed as the Preferred A class still has 50 shares but the Preferred B class now has 75 shares. There are now a total of 125 shares (50 + 75) representing the ownership of this company. The Preferred A class, prior to the warrants execution owned 50%, after the execution of the warrants the Preferred Series A class is reduced to 40% of the company (50/125).

Using the example above, the cost of a failed due diligence can have a meaningful impact on the Preferred A class of stockholders. If the Preferred A stockholders were extremely confident in obtaining milestone #2, they may, for example, offer unusual warrant incentives that would give investors the option to purchase an unusual amount of stock at a lower price if the company *does not* reach the milestone. The warrants expire with the achievement of the milestone. This could bring 50/50 ownership back to the Preferred A class of stock.

Investors typically fund to the next milestone with a little funding left over to allow the company time to raise additional capital for its next round or reserve some cash for a failed milestone. However, investors do not overfund the milestone for a couple of reasons. The first is that they want to ensure that the company does not pursue wasteful activities. Second, the pre-money valuation for the subsequent round does not provide a return on the overinvested monies. As with anything, investors are trying to achieve balance as they lose value if they underachieve a milestone and they lose return if they overinvest in a milestone.

Companies are funded either through an equity investment or through a debt instrument. A debt instrument is simply a promise from the company to repay the debt. These instruments are used in numerous ways to manage risk.

In a young start-up company, it is difficult for the company and the investor to mutually agree on the value of the company. Frequently, an instrument called *convertible debt* is utilized to compensate for this difference in opinion. Although it is not always the case, incubators and angels use convertible debt instruments the most. Unlike the CVC and venture capitalist, incubators and angels frequently do not have sufficient data to set the value for a company. The intention of convertible debt is that a future milestone sets the value for the company. Investors have the option to either get repaid or convert their debt into equity at a future milestone. Although it is generally the lender's decision when to convert or get repaid, some convertible debt instruments may contain mandatory conversion or repayment clauses. These convert or repay clauses are usually set at some future date where the parties agree that the risks associated with the clauses would be removed from the business.

If the company fails to achieve its milestone, convertible debtors will not convert and depending on their terms, may request repayment from the company. This repayment deteriorates the company's cash position. On many occasions, this situation forces the company to go raise additional capital to pay off the debt or the company negotiates terms more favorable to the convertible debt holders. Regardless of the situation, the cash position of the company has been compromised and results in a devaluing of existing stockholder equity.

Not all debt providers are interested in an equity conversion option. Bankers, for example, prefer to make their returns from interest rates on loaned money, not returns on equity. Bankers generally participate when start-ups have meaningful assets for collateral to secure the lending. If the company could not repay the loan, the bank has the right to take ownership of the asset. It would be the intention of the bank to sell the asset to get their return back.

Equity investors expect to have the value of their money grow between investment rounds. Collateral debt financing can offer attractive growth in equity value if used properly. For example, let's assume equity investors expect to have a 20% increase in their stocks' value between each round of equity financing. Continuing, let's assume $25 million is needed to achieve the next value milestone. Equity investors would expect to have $30 million ($25 million plus 20%) in value prior to the start of the next round of financing.

Management could bring even more value to the stockholders by combining equity and debt to fund the next milestone. Continuing, assume that the $25 million was not 100% from equity, instead $20 million was from equity and $5 million was raised as debt. The company would still raise its $25 million but now $20 million is from equity and $5 million is from debt. Assuming the company gets to the next milestone and achieves its planned value of $30 million, the stockholders get $10 million ($30 million – $20 million) in value versus the $5 million ($30 million – $25 million) that was originally planned. Of course, we must subtract the debt interest to determine the net benefit to the company. If the interest rate for the debt was 8% and the milestone took a year to achieve, it would cost the company $400,000 ($5 million × 8%) in interest to bring an incremental $5 million to their stockholders.

The ability for a start-up to obtain the use of convertible or collateral debt instruments rests solely in the start-up's ability to successfully complete the due diligence process that meets the risk profile of the debtor. Debtors have rigorous and conservative due diligence models that has little tolerance for missteps.

DISCUSSION QUESTIONS:

1. Give the reason why a start-up company is designed to be a temporary organization.
2. Define Due diligence and its intent.
3. Give at least two risks associated with due diligence.
4. How is a fundable milestone achieved?
5. What should life sciences companies do to align with the norms of their particular life sciences vertical?
6. How are companies funded?
7. What does equity investment allow in exchange for cash?
8. Provide at least one reason why investors do not overfund the milestone.
9. What does equity investors usually expect during investment rounds?
10. What can management do to bring more values to the stockholders?

NOTES

1 IDEs for Early Feasibility Medical Device Clinical Studies, Including Certain FIH Studies, Guidance for Industry and Food and Drug Administration Staff, Issued October 1, 2013, U.S. Food and Drug Administration, http://www.fda.gov/downloads/medicaldevices/deviceregulationandguidance/guidancedocuments/ucm279103.pdf.

2 Newton's Laws of Motion, *National Aeronautics and Space Administration*, http://www.grc.nasa.gov/WWW/BGH/newton.html (accessed March 15, 2014).

3 Ibid.
4 *Investopedia*, s.v., "Bootstrap," http://www.investopedia.com/terms/b/bootstrap.asp (accessed March 15, 2014).
5 510k Clearances Overview, *U.S. Food and Drug Administration*, http://www.fda.gov/medicaldevices/productsandmedicalprocedures/deviceapprovalsandclearances/510kclearances/default.htm (accessed March 15, 2014).
6 How Do I Know If My Invention Is Patentable? *United States Patent and Trademark Office*, http://www.uspto.gov/inventors/patents.jsp#heading-1 (accessed March 15, 2014).
7 Pre-Market Approval (PMA), *U.S. Food and Drug Administration*, http://www.fda.gov/medicaldevices/deviceregulationandguidance/howtomarketyourdevice/premarketsubmissions/premarket-approvalpma/default.htm (accessed March 15, 2014).
8 *Investopedia*, s.v., "Warrant," http://www.investopedia.com/terms/w/warrant.asp (accessed March 15, 2014).

9 Due Diligence Reputation Is a Critical Business Process

The details of due diligence manifest across all functional positions and aspects of the start-up. Unless its details are identified as a critical business process and integrated into the business process management (BPM) system, its lack of integration unintentionally increases the start-up's risk profile. As a start-up typically is entering uncharted waters, it simply does not seem prudent to start without a map to chart your journey.

The intention of this chapter is to propose a preparation process for due diligence. However, for those start-ups facing a due diligence without the benefit of an integrated process, some methodologies are proposed to achieve the goals of the due diligence.

Those that have time for preparation can be motivated by the implications that an ambiguous process impedes the company's success and could have financial and reputational costs from which it may be hard to recover.

DUE DILIGENCE REPUTATION IS A CRITICAL BRAND COMPONENT

Throughout this book, we discuss the importance of achieving the status of a brand. Using an analogy from a product perspective, the brand is a level of equity built into the product that allows the product to sell a larger volume or at a greater price than it would if it did not possess the intangibles of the brand.

The goal of creating a start-up brand is not dissimilar to that of a product. A start-up should strive to create a reputation that allows the company to be valued more than the sum of its parts (It is also the responsibility of the company to create a brand for its product.) A well-planned and managed due diligence process reduces the risks inherent to the process itself. A well-executed due diligence process with consistent messaging flushes out the risk and leaves the reward of reputation.

A competently run due diligence process will result in more frequent deal consummation. Even a failed deal should leave the other party with a favorable impression: an impression that is frequently shared among industry insiders.

DO WHAT YOU SAY, THEN THEY WILL TRUST WHAT YOU FORECAST

I have a saying associated with due diligence: *If you do what you say, then they will trust what you forecast*. Although the message of this book and chapter is to

DOI: 10.1201/9780367533052-9

proactively plan for due diligence from the company's very formation, the author appreciates that one may learn this lesson midstream.

As life science start-ups can take 5–12 years to exit, they rarely begin with the thought of creating a due diligence process. During their first few years of life, companies tend to focus on initial funding, patents, and generating proof of concept. This focus can leave documentation gaps and missing details that will be requested upon a company's first major due diligence.

The due diligence process should be part of a company's overall BPM system. BPM's goal is to align all of the organization's business processes to ensure that the company's goals and objectives are met.[1] Business processes are the *means* or methods used to achieve the *ends*, which are the goals and objectives.

For the start-up planning its first major due diligence and finding itself in a situation with documentation gaps and missing details, there is a reasonable strategy that can be deployed to provide confidence to the party performing the due diligence. The start-up needs to recognize that the missing details are frequently means-oriented (how it was done) and if the company can demonstrate that the ends were met (the goals or objectives), there can be a reasonable assumption that the means were appropriate. In situations where the means are important but you are lacking the internal details, provide confidence from another independent organization. Let's expand upon these two concepts.

If the company can demonstrate that the ends were met, then one may be able to avoid a means discussion. If a means discussion cannot be avoided, then try borrowing the reputation of another party who has a means reputation.

For example, if a patent has been issued to the start-up from the U.S. Patent and Trademark Office (USPTO), clearly one can assume that there was quality in the preparation process. If a patent has not been issued, and is either in development or at a patent pending phase, the only confidence one can provide is to use the reputation of the preparer by using a nationally recognized law firm. The due diligence organization can assume competence if a branded law firm is preparing the patent.

A problem frequently occurs when the company hires an individual or small firm to prepare the patent. The start-up's motivation is to save cost. This does not necessarily suggest that the individual is not of quality because frequently they were former employees of the branded firm. Unfortunately, even the highest quality individual seldom has the national reputation of their former firm. As a result, when due diligence is performed, the organization performing the due diligence assigns a greater risk to the ability to obtain a patent than they would if the branded firm prepared the patent. This is a difficult situation for the start-up to overcome because the assigned risk is the opinion of the due diligence firm. Once the risk is assigned, it is difficult to change the due diligence organization's opinion.

I have observed seasoned entrepreneurs proactively manage this situation by using talented, cost-effective individuals to perform the analysis and prepare the patent draft. The start-up later transfers the patent portfolio to a branded firm for review and submittal prior to due diligence. The benefit appears twofold: the patent was prepared cost-effectively and yet represented by the branded firm to the due diligence organization. There is also a third benefit to this strategy. In larger firms, the detailed work is generally relegated to lower-level associates. These lower-level associates will assuredly be competent but they generally lack the 15-plus years experience of the independent. Given the independent's experience, they may be able to uncover a creative approach that does not readily avail itself to the lesser-experienced associate.

As stated previously in Chapter 8, the legal intent of due diligence is to demonstrate that a level of prudence was taken to validate what you thought you were getting in a relationship. From the perspective of the organization performing the due diligence, their desire is to demonstrate this level of prudence in the fastest and most inexpensive way possible. This is a motivation also shared by the start-up.

One of the fastest and least inexpensive ways of completing a due diligence is not only by preparation but also by utilizing others to build a body of evidence. Not unlike a court trial, the goal would be to utilize leverage historical interactions as proof sources. The author has created a "Due Diligence Checklist" in Excel that can be downloaded from the book's Web site, at: https://healthcaredata.center/commercialization-startups; the spreadsheet lists some potential end and means validation points that a company may utilize to create a case of validation.

AMBIGUOUS PROCESSES ARE A QUALITY ASSURANCE PROBLEM

Many of us played sports when we were children, and most likely our first act was learning the basic skills necessary to play the game, such as how to throw or catch a baseball, kick a soccer ball, or throw a basketball through the hoop. To play the game and determine a winner, however, basic skills are not enough. We needed to be taught to play by the same rules so the final score or outcome had meaning. For example, if there were no rules in soccer, one could simply run with the soccer ball down the field and throw it into the net while another team kicked the ball down the field and kicked it into the net. The final score of the game would not be a meaningful measurement of which was the better team given the different methods each team used.

Without rules, or standards, products and services may not work as expected and they could not be compared. For example, if one was to buy a light bulb and the package stated that each light bulb lasted 200 hours, you may choose to buy the cheaper light bulb. What if the cheaper light bulb lost 1 footcandle (of illuminance) every 50 hours it was used? Would this not be something you would deem important in your decision? To ensure adequate comparison, standards and measurements are necessarily adopted by businesses to provide confidence that everyone is playing by the same rules. To achieve this confidence, numerous product categories and company external reporting and funding standards exist. Adopting these standards and demonstrating compliance to them gives confidence in a due diligence proceeding. This confidence is achieved by having an integrated quality assurance and quality control program that organizes all the subsystems for the entire business.

What do we mean by quality assurance? The American Society of Quality Assurance (ASQ) defines assurance as the act of giving confidence. Quality assurance is a set of planned processes and systemic activities implemented in a system so that quality requirements for a product or service are fulfilled.[2] Control is an evaluation to indicate the need for a corrective response. Thus, quality control is the techniques and activities deployed to fulfill the requirements set forth in the quality assurance system.[3] The question to the start-up is what is the overall system that ensures you are managing and prepared for a due diligence exit?

A start-up company is constantly under pressure to achieve that next value or fundable milestone. This laser focus combined with limited personnel and funding frequently results in an ambiguous due diligence process or one that is seen as an *event*. Yet, this same start-up company is most likely well aware of the consequences of their products not being designed and manufactured according to industry or regulatory standards. They are most likely aware that their accounting and tax techniques must comply with specific standards. They are most likely aware that their fund-raising activities must comply with Security and Exchange Commission (SEC) regulations. Some of the better-known regulations are provided below:

- *ISO Standards*: Globally, the International Organization for Standardization provides guidance for products, services, and good practices. The ISO is an independent, nongovernmental organization made up of members from the national standards bodies of 163 countries.[4] There are several ISO standards that are specific to the life sciences industry.
- *CE Mark*: In Europe, there is a conformity marking called the *CE Marking*, which is the declaration that the product meets EU required directives.[5] All pharmaceutical, diagnostics, medical devices, and equipment require CE Marking before they can be sold.
- *cGMP Compliance*: In the United States, the FDA augments traditional quality assurance and control systems with cGMP. cGMP refers to Current Good Manufacturing Practices that assure proper design, monitoring, and control of manufacturing processes and facilities.[6]
- *FDA Approval*: The U.S. Food and Drug Administration (FDA) must approve most life science products before they can be sold in the United States. Its approval demonstrates that the company has provided data to convince the FDA that the product's benefits outweigh its risks.[7]
- *GAAP Reporting*: The Federal Accounting Standards Advisory Board (FASAB) sponsors efforts to improve federal financial reporting through a hierarchy of generally accepted accounting principles (GAAP).[8] External financial statements and reporting therefore must meet these standards.
- *SEC Regulation*: The mission of the SEC is to protect investors, maintain orderly and efficient markets, and facilitate capital formation.[9] Specifically, Regulation D is an exemption offered by the SEC to allow for the sale or equity of debt securities to investors seeking to raise private capital.[10] In order to raise private equity these standards must be met.

Ambiguity is something that does not have a single meaning and by its very nature increases risk. Quality assurance requires defined, planned, and systemic activities.

Its supporting quality control techniques ensure that when something is ambiguous, it is subsequently defined as part of a corrective action system. With all the standards and regulations that a start-up company adopts, why is it that there is not an industry standard for start-up due diligence?

Standards and regulations do not come about without industry or regulators investing significant money and time. Because start-ups are formed with the intention of having an exit within a fairly short period of time while big and small businesses are formed with long-term intentions, standards are rarely created for start-ups as they are not designed to be around long enough to influence them. Due to their intention to exit, life science start-ups must design their own due diligence processes by combining components of known standards with norms that are specific to achieving an exit in their particular industry vertical. These norms are frequently not associated with explicit industry standards but defined by the vertical's executives (buyers) and investors (start-up funders).

THE COMPONENTS OF A DUE DILIGENCE BUSINESS PROCESS

Thus far in our conversation, the reasons for due diligence and their associated risks have been detailed. Start-up organizations must constantly expand their funding and their relationships to achieve their ultimate goal of a successful exit.

Sir Isaac Newton's laws of inertia underline the importance of meeting value and fundable milestones. His law of acceleration demonstrates the necessity of increasing resources as the company progresses to market. The closer to commercialization, the more market forces push against the start-up. The company must prepare to possess the resources to provide a force greater than the resisting forces of the market.

The fragility of reputation warrants the utmost attention to brand messages. A start-up cannot be perfect in every encounter and occasionally a bad interaction will happen. The story of Bill at the J.P. Morgan meeting (Chapter 8) reminds us that unintentional messages can slip into the marketplace. The strength of expanding positive relationships and the achievement of a milestone can be the best defense to thwart a negative encounter.

Not all start-ups will have the resources, talent, funding, or time to prepare for due diligence. In those scenarios, strategies and tactics have been provided to make the best of the situation.

Of course, the hope for this book is that it is recognized that preparation for due diligence is not only a brand strategy but also a tactical business process that must be included in your BPM plans. This leaves us with the final topic of due diligence—how do you integrate the strategic and tactical components into one comprehensive system?

Strategic due diligence must start with the start-up company's investor pitch (A PowerPoint template can be found on the book's Web site, at: https://healthcaredata.center/commercialization-startups) Nobody has expressed this better than Stephen Covey's concept of *begin with the end in mind.*[11] The investor pitch should outline the investment premise for investors in the organization. The investor pitch should also communicate the start-up's exit strategy and answer the question of why someone would want to buy the start-up and who.

The start-up would want to be able to document the following strategic due diligence messages. These messages will ultimately be the brand messages to the due diligence constituencies:

- The start-up has identified a clear problem in the market.
- The start-up possesses the technology and/or know-how to win.
- The start-up has domain knowledge as expressed through their personnel and relationships.
- The start-up has the capability to access the market.
- The start-up must articulate who will buy them and why.
- The start-up must express the milestones (value and fundable) of the journey. They must demonstrate that these milestones meet industry norms.
- The start-up must express the time and cost of the journey. The start-up must demonstrate that their costs and time frames align to industry norms.
- The start-up must express the fundable milestones and the expected exit points and valuations.

Tactical due diligence starts with the start-up company's business plan. A business plan's goal is to define your business and its goals and expresses how you plan to achieve a return for your investors. It details the company's major tactics, their organizational components, and how resources are allocated. It speaks to the risks and how the start-up would plan to handle them. The business plan concludes with your financial statement in the form of a pro-forma balance sheet, an income statement, and a cash-flow statement.[12]

The business plan articulates the start-up's products and services. The company must be prepared in their due diligence preparation to demonstrate that the technical specifications of their product provide the competitive advantages expected.

The company must demonstrate that they have knowledge of their market as expressed through a marketing plan. The marketing plan should demonstrate the company's knowledge of the industry as demonstrated through primary and secondary research. Primary research is the gathering of data by the company itself and secondary research utilizes public information. When a company is young, secondary research is frequently given more creditability by due diligence organizations as this information represents the view of independent third parties. It is felt to be unbiased. As the company has more of a presence in the marketplace as demonstrated by sales force size and/or revenue, primary research is given more credibility.

The marketing plan needs to detail the economics of the industry, the size of the market, and the expected market share. It details the target market and the potential growth opportunities as well as the market barriers.

Life sciences operate within an environment of great complexity as its disease states and its value chain are complex. Components of an appropriate detailed marketing plan include the following:

- The objectives and critical success factors of the plan
- The market overview details that are specific to life sciences such as:
  - Disease prevalence and incidence rates
  - Diagnostic strategies to increase the identification of patients
  - Treatment types and the percentage of those types as it relates to diagnosed patients
  - The units per treatment
  - The average selling prices in the market
- The health care market environment including:
  - Buyer segmentation inclusive of group purchasing organization and integrated delivery networks
  - Physician overview (types of physicians)
  - Procedural overview
  - Hospital overview (segmentation, consolidations, bankruptcies)
  - Payer overview
  - Hospital and physician P&L trends
  - Diagnostic technology trends
  - Therapeutic technology trends
- The patient profile:
  - Disease prevalence
  - Procedure/therapeutic segmentation
  - Patient referral sources
- The company's distribution and sales strategy
- The company's market development strategies
- The company's communication strategies
- Competitor analysis:
  - Market growth by competitor
  - Competitor presence in segments
  - Technology platform evaluation
  - Strengths, weaknesses, opportunities, and threats (SWOT) analysis
- Relative market strength evaluation

The company must detail its operational plan to support the organization. The start-up must explain the operations of the business and the external standards that the organization must meet to be in compliance with the numerous regulations in the life sciences marketplace. Details around production techniques, personnel, key suppliers, costs, quality, and inventory methods must be provided.

The company must detail its management and organizational components, inclusive of headcount and its corresponding timing. An organizational chart should be presented to articulate the structure of the company.

The business plan ends with multiyear financial statements. These statements summarize the business plans in terms that investors can grasp, as it relates to understanding funding and the projected return on investment.

DUE DILIGENCE CHECKLIST

	Strategic & Tactical Systems						Balance Sheet			INCOME STATEMENT			
	Mkt/Strategic Plan (3–5 yrs)	Partnerships & Brand	Annual Business Plan	Performance & Process Mgmt	Legal, Risk, & Regulation	Personnel	Assets	Liabilities	Equity & Retained Earnings	Revenue	CGS	GM	Expenses & IDAT
• Understand any claims made since inception by any person asserting a right to an equity interest in the company									✓				
• Identify any restrictions on the company's ability to make dividend payments									✓				
Employee Matters													
• Employee policies and handbook				✓									
• Compensation agreements						✓							✓
• Compensation levels				✓									
• Labor contracts						✓					✓		✓
• Bonus and profit sharing agreements									✓				
• Retirement plans; defined benefits, thrift, pension, profit sharing						✓							✓
• Post-employment agreements; salary continuation agreements						✓							✓
• Union-sponsored multiemployer plans						✓					✓		✓
• Stock bonus plans									✓				
• Employee stock ownership plans									✓				
• Medical plans						✓							✓
• Dental plans					✓	✓							✓
• Short-term disability or sick plan arrangements						✓							✓
• Long-term disability insurance or uninsured arrangements					✓	✓							✓
• Group term or other life or accident insurance						✓							✓
• Unemployment benefits						✓							✓
• Vacation benefits						✓							✓
• Welfare plans						✓							✓
• Director or officer deferred fee plans													✓
• Excess benefits plans (excess of IRS code limitations)						✓							✓
• Stock option, purchase or stock bonus plans						✓							✓
Financial Matters													
• Accounting policies													
• Leases; equipment, vehicles, etc.							✓						
• Accounting methodology changes													
• Auditing policies and work papers													
• Contracts greater than 1 year or $xx								✓					✓
• Review of any transaction between company and any shareholders, member, officer, director, employee, or affiliates													
• Logistic or warehouse contracts							✓				✓		
• License agreements							✓						✓
• Franchise agreements										✓	✓		
• Joint venture agreements	✓	✓							✓				
• Partnership agreements		✓							✓				
• Distribution agreements		✓								✓	✓		
• Joint marketing agreements		✓								✓			✓
• Purchasing agreements			✓				✓						
• Professional service agreements			✓								✓		✓
• Capital expenditure agreements		✓	✓										
• Contracts for sale of assets (noninventory)							✓						
• Judicial and administrative orders					✓								✓

FIGURE 9.1 (*Continued*) Due Diligence Checklist (a).

DUE DILIGENCE CHECKLIST

	Strategic & Tactical Systems						Balance Sheet			INCOME STATEMENT			
	Mkt/Strategic Plan (3–5 yrs)	Partnerships & Brand	Annual Business Plan	Performance & Process Mgmt	Legal, Risk, & Regulation	Personnel	Assets	Liabilities	Equity & Retained Earnings	Revenue	CGS	GM	Expenses & IDAT
• Building and construction agreements		✓	✓				✓						
• Lease agreements for machinery, equipment, or property							✓				✓		✓
• Computer contracts							✓						✓
• Contracts with government entities such as NIH, DoD, etc.										✓			
• Debt agreements: notes, mortgages, leases, license, franchise, or covenants that would be violated or triggered by an acquisition								✓					
• List of bank accounts				✓									
• Review lists where company is identified as an approved vendor		✓		✓						✓			
• List of awards, etc.		✓											
• Review major transactions between the company and its subsidiaries	✓	✓		✓						✓	✓		✓
Financial Performance													
• Accounts receivable aging							✓						
• Accounts payable aging								✓					
• Major growth drivers and prospects										✓			
• Predictability of business	✓									✓	✓	✓	✓
• Risks to foreign operations (exchange rate fluctuations, government instability)	✓		✓		✓								
• Industry and company pricing policies				✓									
• Income statements, balance sheets, cash flows, and footnotes					✓		✓	✓	✓	✓	✓	✓	✓
• Planned versus actual results				✓	✓								
• Management financial reports				✓									
• Product type (Sales and GM)					✓					✓		✓	
• Channel (Sales and GM)										✓		✓	
• Geography (Sales and GM)										✓		✓	
• Identify unusual trends to identify nonrecurring income or expense items										✓			✓
• Revenue recognition policy				✓						✓			
• Accounting treatment of sales discounts				✓						✓			
• Concentration of customer risks										✓			
• Backlog				✓	✓					✓			
• Bank accounts and depository arrangements							✓						
• Inventory accounting policy				✓	✓		✓				✓		
• Trend expense categories: controllable, noncontrollable	✓			✓	✓						✓		✓
• Instruments that could create liens					✓								
• Receivable analysis							✓						
• Payable analysis								✓					
• Licensing agreements for nontangible assets	✓									✓			
Financial Projections					✓								
• Revenue (by region, by customer, product line, etc.)	✓				✓					✓			
• Gross margin	✓											✓	
• EBITDA					✓								✓
• Working capital													
• Economic assumptions underlying projections	✓				✓								
• Explanation of projected capital expenditures, depreciation and working capital arrangements	✓				✓		✓						✓
• External financing assumptions	✓							✓	✓				

FIGURE 9.1 (*Continued*) Due Diligence Checklist (b).

DUE DILIGENCE CHECKLIST	Strategic & Tactical Systems						Balance Sheet			INCOME STATEMENT			
	Mkt/Strategic Plan (3–5 yrs)	Partnerships & Brand	Annual Business Plan	Performance & Process Mgmt	Legal, Risk, & Regulation	Personnel	Assets	Liabilities	Equity & Retained Earnings	Revenue	CGS	GM	Expenses & IDAT
Tax Information													
• Federal, state, local, and foreign tax returns													✓
• Audit reports of tax authorities and any open issues					✓								
• Real estate tax bills and payment records							✓	✓					
• Personal property tax bills and payment record								✓					
• Communications between tax authority and company													
• Franchise, license, capital stock, doing business, and similar tax reports													
• Loss carryforward													✓
Management and Operations													
• Understand significant third-party vendors		✓			✓						✓		✓
• Proprietary information	✓				✓		✓						
• Patents and patent applications	✓				✓		✓						
• Copyrights	✓				✓		✓						
• Organizational chart				✓									
• Noncompetition agreements					✓	✓							✓
• Trademarks, service marks, logos, trade names	✓				✓		✓						
• Historical and projected headcount by function and location	✓			✓									
• Nonpatentable proprietary know-how (trade secrets)	✓				✓		✓						
• Biographies of key personnel						✓							
• Employee relations issues						✓							
• Personnel turnover				✓		✓							
Real Estate and Equipment and Other Personal Property													
• List of owed items							✓						
• List of sold items							✓						
• List of leased items					✓		✓						
• Appraisal values to book value							✓						
Insurance													
• Fire					✓								
• Liability					✓								
• Casualty					✓								
• Life					✓								
• Title					✓								
• Worker compensation					✓								
• Director and officer liability					✓								
• Claim and loss history, correspondence with carriers					✓								
Government Regulation													
• Licenses, permits, and filings required or made with any government agency													✓
• Correspondence with regulatory authorities					✓								
• Accident or injury reports with filings				✓	✓	✓							✓
Litigation and Claims													
• Pending or threatened litigation, regulatory investigations, governmental actions, arbitrations, or notices of violation					✓								✓
• Files and records of past litigations					✓								
• Environmental and safety issues					✓								✓
• New or pending regulations and their consequences	✓				✓					✓	✓		✓
• History with SEC, FDA, EPA, and other regulatory agencies	✓				✓								

FIGURE 9.1 (*Continued*) Due Diligence Checklist (c).

DUE DILIGENCE CHECKLIST

	Strategic & Tactical Systems						Balance Sheet			INCOME STATEMENT			
	Mkt/Strategic Plan (3–5 yrs)	Partnerships & Brand	Annual Business Plan	Performance & Process Mgmt	Legal, Risk, & Regulation	Personnel	Assets	Liabilities	Equity & Retained Earnings	Revenue	CGS	GM	Expenses & IDAT
Research & Development													
• Capabilities	✓												
• Strategy	✓												
• Key personnel	✓					✓							
• Major activities	✓				✓								
• New product pipeline	✓												
• Status and timing	✓												
• Cost of development	✓												
• Critical technologies necessary to implement risks	✓												
Marketing, Sales, and Distribution	✓												
• Domestic and international distribution channels													
• Positioning of company and its products	✓												
• SWOT by segment	✓												
• Customer pipeline analysis	✓												
• Existing customer status and trend of relationship	✓			✓									
• Sales force	✓									✓			
• Compensation	✓												
• Quota	✓												
• Sales cycle	✓												
• Expansion plans	✓												
Products	✓												
• Major customers and applications													
• Historical and projected growth rates	✓									✓			
• Market share	✓									✓			
• Speed and nature of technology change	✓									✓			
• Timing of new product launches	✓				✓					✓			✓
• Cost structure and profitability	✓									✓			✓
Customer Information	✓						✓				✓		
• List of top customers inclusive of revenue													
• List of strategic relationships	✓				✓					✓			
• List of significant customers lost and reasons							✓		✓				
• List of top suppliers										✓			
Competitive Information							✓				✓		
• Market position and related strengths and weaknesses in each market segment													
• Basis of competition by segment (price, service, technology, distribution)	✓									✓			
	✓									✓			

FIGURE 9.1 Due Diligence Checklist (d).

A discussion of due diligence is not complete without proposing its integration into the company's overall BPM systems. A "Due Diligence Checklist," Figure 9.1, has been created so that the due diligence process can link into key business management processes. This is necessitated as the requirements of an investor's or acquirer's diligence may require more evidence or impose constraints or additional data requirements than an internally focused process may require. To assist the reader in connecting their business plan to their BPM, a "Due Diligence Checklist" is available for download from the book's Web site, at: https://healthcaredata.center/commercialization-startups/due-diligencechecklist/.

DISCUSSION QUESTIONS:

1. Explain the effects of a competently run due diligence process.
2. What is the goal of BPM?
3. Why should the due diligence process be part of a company's BPM system?
4. Explain the benefit(s) of transferring the patent portfolio to a branded firm for review and submittal prior to due diligence.
5. What does the ISO provide?
6. What does a marketing plan contain?
7. Why does a company need an operational plan?

NOTES

1 Business Process Management, *Tech Target*, January 3, 2011, http://searchcio.techtarget.com/definition/business-process-management.

2 American Society of Quality Control, Quality Assurance and Quality Control, *ASQ*, http://asq.org/learn-about-quality/quality-assurance-quality-control/overview/overview.html (accessed March 15, 2014).

3 Ibid.

4 What Is ISO? *International Standard Organization*, http://www.iso.org/iso/home/about.htm (accessed March 15, 2014).

5 CE Marking, CE Marking—Enterprise and Industry, *European Commission.*

6 Facts about Current Good Manufacturing Practices, *U.S. Food and Drug Administration*, http://www.fda.gov/drugs/developmentapprovalprocess/manufacturing/ucm169105.htm (accessed March 15, 2014).

7 What Does "FDA Approval" Mean? *Rottenstein Law Group*, http://www.rotlaw.com/legal-library/what-does-fda-approvalmean/ (accessed March 15, 2014).

8 FASAB Mission, The Mission Supports Public Accountability, *Federal Accounting Standards Accounting Board*, http://www.fasab.gov/about/mission-objectives/ (accessed March 15, 2014).

9 The Investor's Advocate, Introduction, *U.S. Securities and Exchange Commission*, http://www.sec.gov/about/whatwedo.shtml (accessed March 15, 2014).

10 Regulation D Offerings, *U.S. Securities and Exchange Commission*, http://www.sec.gov/answers/regd.htm (accessed March 15, 2014).

11 Stephen Covey, *Seven Habits of Highly Effective People* (New York: Simon & Schuster, 1989).

12 What Is a Business Plan and Why Do I Need One? *Small Business Administration*, http://www.sba.gov/content/what-business-plan-and-why-do-i-need-one (accessed March 15, 2014).

Section III

Align with the Industry Norms

10 Find the Industry Norms

The best way to start this discussion is to share a formative experience. As an executive-in-residence, I was working with a company called *Blue Belt Technologies* and their founding CEO Craig Markovitz. Blue Belt Technology has the ability to incorporate surgical planning and navigation with a hand-held robotic-assisted bone preparation device. The technology was developed at the Robotics Institute at Carnegie Mellon University with commercial development directed by company cofounder Branislav Jaramaz, Ph.D. From the very beginning, the company knew its technology would be very impactful on hip, knee, and shoulder surgery.

The company's concern in approaching these markets was that the revenue size and sales channel control of the large orthopedic companies such as Johnson & Johnson, Stryker, Zimmer, Biomet, Medtronic, and Smith & Nephew could inhibit access to customers. Access to the larger funding amounts necessary to enter the hip, knee, and shoulder surgery markets was also of concern as the company was based in Pittsburgh and did not have numerous regional venture capital organizations to support it.

Given the above, the company sought a procedure where they could have a clinical impact and the funding to get to market was not as great a concern. The company targeted the clinical problem of spinal restenosis. Spinal stenosis is a narrowing of the spinal canal. This narrowing restricts (pinches nerves) in the spinal canal causing pain, numbness, and loss of motor control. The surgical resolution for this problem is a procedure called *Lumbar Decompressive Laminectomy*. This procedure improves the limited space by removing bone and soft tissue in order to decompress the nerves.[1] In a portion of these procedures, the stenosis returns post procedure—this is called *spinal restenosis*. Blue Belt's initial thought was that this would be a great entry procedure for the company, as the impediments and cost to market would be modest. For the physician whose first procedure removed the majority of the suspect bone and tissue, the second procedure required improved precision. Blue Belt's technology offered that precision and the funding to enter that space was more obtainable in the midwest funding markets. The thought was that success in this segment would validate the technology in the marketplace. This validation could lead to an exit or achieve a fundable milestone allowing for entry into the venture capital class to fund the bigger hip, knee, and shoulder segments.

The eventual clarity of the above strategy was not without its vacillation. Subsequent to this decision, I had an opportunity to present the company's product to Daniel Cole, who is a general partner at Spray Venture Partners. Dan was a group president of the $1.2 billion vascular business when I was at Boston Scientific. He has served on more than 23 med-tech boards of directors and has the ability to immediately see through complex issues with great clarity.[2] I showed him how the

DOI: 10.1201/9780367533052-10

technology worked and Dan was impressed. He then asked, "What are you going to do with it?" I shared that we were targeting spinal restenosis. I will never forget Dan's comment: "This is an important technology chasing a relatively unimportant clinical issue." The clarity of Dan's comment was devastating and he was kind enough to offer additional opinions on where to take this technology that would be of interest to him.

Returning to Pittsburgh and discussing the feedback with the company CEO, it eventually dawned on us the importance of context as it relates to Dan's comments. Dan's feedback was from the perspective of a venture capital investor; he was not suggesting that restenosis was an unimportant clinical issue to the physician or patient. What he was suggesting is that this clinical target did not have enough procedures to provide a return for venture capital investors.

This too needs to be placed in context. Venture capital funds come in various sizes, such as funds under $25 million, funds under $50 million, funds under $250 million, and funds as large as $1 billion. These funds have a limited number of management personnel and it is the personnel that determines the number of investments that can be made, monitored, and managed. Let's discuss this further: assume a fund has three fund managers or venture partners and each are capable of managing three or five companies at a time. For simplicity let's assume each manager is capable of managing four companies for a capacity of 12 companies (4 companies × 3 managers). The venture firm, understanding their limit of 12 companies, would look to uncover opportunities in a manner that matches their fund. For example, if the fund had $100 million under management, the company would be looking to deploy $8.3 million per company. If a company needed $2.5 million to achieve an exit, it is likely that a venture capital fund of $100 million would pass on the opportunity because it is not big enough.

Over the course of the company's development, Craig had forged a congenial relationship with the large orthopedic companies. They appeared very interested in the technology but the technology was too early (contained too much risk) for them to acquire it. However, these discussions provided Craig with the confidence that if he could get the technology de-risked, he had an exit. Using Figure 10.1, let's take an inventory of where the company was at this point.

The company clearly had significant validation that their technology was important. Their Small Business Innovation Research (SBIR) was an independent validation of the technology's importance. The company had created an IP Pyramid analysis to demonstrate that they deeply understood all of their individual IP properties and how those properties interacted with each other and more importantly, future acquirers. (We will demonstrate how to create an IP Pyramid in Chapter 16.) Last, both respected venture capitalists and acquirers validated that the technology was important.

Blue Belt Technologies third founder is Dr. Anthony Digioia, III, who is an orthopedic surgeon. He is well known in the industry among his peers and the manufacturers (interested acquirers). This fact also de-risked the company as it gave confidence that the product's specifications would be aligned to industry clinical norms. Independent physicians had also validated that the procedures the company sought were clinically important.

Technology Validated as Important	Interested Acquirers
• SBIRs validate technology by independent third-party thought leaders • I.P. Pyrimid complete • Patents filed • Respected VC states technology is important • Acquirers inquiring about the technology	• Two major orthopedic companies express long-term interest

Funding	Clinical Customers
• SBIR nondilutive	• Company founded by an orthopedic surgeon
• Incubator funding — PLSG • Angel or Super Angel — T-B-D • Venture capital — too early	• Individual physician feedback • Spine procedure validated • Knee procedure validated
• Corporate VC — unknown	

FIGURE 10.1 Blue Belt Technologies evolution of relationships.

Skipping the *Funding* column in the table for a moment, Dr. Digioia's participation alone piqued the interest of future acquirers because they knew him as an innovator. The CEO had utilized his relationships with manufacturers to keep them informed enough to be intrigued but uninformed enough to protect his intellectual property from replication before his patents were issued.

Finally, the company's funding strategy also demonstrated an effort to communicate validation and de-risking to downstream funders. The SBIR funding was non-dilutive and this de-risked the company for the next investor class. For example, if it is expected the company should take \$10 to get to a value milestone and the company gets there for \$6 because they used non-dilutive funding versus equity funding, the next investor sees \$4 of future risk that could be covered. Continuing, if the second value milestone is at \$20 and the company gets there for \$24, they did not lose equity value as the \$4 overage is covered by the \$4 of non-dilutive SBIR funding. If the company gets there as planned, there is still \$4 of value to be obtained by the stockholders. Last, the company had received an investment from an incubator, in this case, the Pittsburgh Life Sciences Greenhouse (PLSG). The PLSG incubator performs a due diligence before they invest so the next investor has confidence that the company can withstand a due diligence.

So what did the company do next? The company had received feedback from the venture capital industry that they needed to move into a bigger procedure than spine. However, if the company did that, the fundable milestone to achieve entry into the VC funding class was millions of dollars away. The company also recognized that the large manufacturers who had expressed interest also had corporate venture capital funds. If the company picked the spine procedure, they could seek angels and potentially bridge to corporate venture should that even be necessary.

The company pursued this strategy and found an angel. Subsequently, to the company's surprise, the angel was actually a super angel. We previously spoke about the value of a super angel, not only in terms of funding, but in their expertise and industry relationships. On December 8, 2011, the company announced that Healthpoint Capital acquired the company for an undisclosed amount. Healthpoint commented in their announcement that this was their "first acquisition to create a world-wide leader in providing a continuum of care for patients suffering from orthopedic diseases."[3]

In the end, a private equity firm saw the same long-term potential that the venture capital firms saw. However, as stated in the press release, the private equity firm was seeking to build something bigger than one company. A venture capital firm would be determining return based upon the individual investment property. By targeting a modest procedure and through a little luck obtaining a super angel, the company was able to get their successful exit. The exit just was not in the manner they originally envisioned. However, by continually building value through non-dilutive funding, obtaining independent third-party validation of the technologies, and by understanding the context of their funding marketplace (Pittsburgh), the company plotted a pathway to create a successful exit.

What is implicit in this story is that the company was trying to align the needs of the multiple constituencies. For the health care system, the company was seeking to solve an important problem. Simultaneously, the company was trying to raise capital to fund the venture. In its process of seeking funding, they uncovered the financing goals of the various players: angels, venture capital, and corporate venture. They aligned their development plans within the industry's norms and uncovered the purchase triggers of their various acquirers. In the end, this company was sold to Smith & Nephew for $275 million in 2015.

DISCUSSION QUESTIONS:

1. What was Blue Belt's rationale for their initial approach to commercialization?
2. How did the venture capital world see this approach as flawed?
3. What was Daniel Cole suggesting regarding the statement: "This is an important technology chasing a relatively unimportant clinical issue"?

NOTES

1 John L. Zeller and Cassio Lynm, Spinal Stenosis, *The Journal of the American Medical Association*, February 27, 2008, http://jama.jamanetwork.com/article.aspx?articleid=181530.
2 Daniel Cole, Biography, *Spray Venture Partners*, http://www.spraypartners.com/team_cole.html (accessed March 15, 2014).
3 News Events, *Blue Belt Technologies*, December 8, 2011, http://www.bluebelttech.com/december-8-2011/.

11 Solve an Important Customer Problem

Innovation, as noted previously, is defined as introducing something new to affect change. From a business perspective, a marketing specialist would state that the desired effect of innovation is either to:

- Create a new, differentiated, and protectable market category, or
- Collapse the value steps or improve efficiency in an existing category, resulting in decreased cost or an increased benefit.

Our previous discussion on Blue Belt's technology has both components of innovation that would excite a marketeer. Ultimately the company has an opportunity to create an entirely different category as their bone-sparing technology could allow for new implant shapes. Today's technologies require so much bone removal that patients generally are capable of only receiving one artificial hip or artificial knee in a lifetime. The potential of this technology could be that individuals could receive more than one artificial implant in a lifetime. In the short term, the company was addressing the problem of spinal restenosis where precision in the hands of a skilled surgeon would be valuable in reducing complications.

Although both the long-term approach and the short-term approach provide value, how can we tie this to the comment made by the venture capitalist: "This is an important technology, chasing a relatively unimportant procedure." We must again restate that this was not a comment to the unimportance of spinal restenosis, it was intended to represent the size of the market. How can we put this into context? How can we create an analytical model that allows us to visualize this comment?

The author uses a spreadsheet that he calls a *Disease State Fact Book.* Figure 11.1 represents what one would look like.

Let's walk through this so that you understand its value. The data provided in the example is for the disease state of coronary artery disease. The information is historical and intentionally out of date as to not contradict existing companies' projections for these markets.

Starting at the top on row 1, we have *Disease Prevalence*, which is defined as the portion of the population projected to have the specific condition. Disease prevalence is derived via public or population health research and then statistically applied to population numbers to obtain the number of prevalent patients. This can then be applied to country or regional populations to forecast specific conditions. Prevalence does not necessarily translate into someone needing a procedure in a given year. In our example of coronary artery disease, someone in their 30s could have evidence of the disease, however, the disease may not have progressed far enough to present clinically or become diagnosed.

DOI: 10.1201/9780367533052-11

Disease State Fact Book

Row #			Base Year	Year 2
1	Disease Prevalence	Portion of the population found to have the condition (1 in 1,000)	24,652,555	25,268,869
2	Incidence %	Percentage of new cases (generally a year)		20%
3	Incidence	Occurrence of new cases since last time period later year or in a period of time (generally a year)		616,314
4	Percentage Recurring	Percentage of population with a recurring event in a given year		43%
5	Prevelance Population	[Disease prevalence less incidence] × percentage recurring		10,812,795
6	Number Diagnosed	Number diagnosed patients (the act of identifying treatable disease)		11,429,109
7	Diagnosis Rate %	Number diagnosed/disease prevalence (this included incident patients)		45.2%

Row #			Medical Therapy	CABG	Interventional Procedure
8	Procedural Approaches	Diagnostic, Medical Devices, Pharmaceutical, Long-Term Care, Rehabilitation, etc.			
9	Procedure/Service Approach %	The percent of diagnosed cases that would use this product/service	84.5%	3.5%	12.3%
10	Number of Procedures/Services	Number of diagnosed x procedure/service approach %	9,658,740	400,019	1,406,923

Row #			Stent	Guide Catheter	Guide Wire
11	Type of Products/Subservices	List the individual products or services performed			
12	Units per Procedure/Service	Example: 2 stents per procedure, 30 pills per cycle, 30 days in long-term care	2.2	1.75	1.1
13	Market Units/Services	Number of Procedures × Units per Procedure/Service	3,095,231	2,462,116	1,547,616
14	Average Revenue per Event	Revenue value per event or service — note revenue by manufacturer would be different than at the hospital level	$ 112.23	$ 9.87	$ 6.93
15	Market Dollars or Cost	Market Units × Average Price	$ 347,377,822	$ 24,301,084	$ 10,724,977

FIGURE 11.1 Disease State Fact Book.

Incidence Percentage and *Incidence* are in rows 2 and 3. Incidence is the number of new cases in a given period of time, generally a year. There are various ways of calculating this. The most theoretically correct is to uncover research associated with new cases. As these individuals are diagnosed, medical records provide an accurate accounting. In building the model, some individuals put rigor into accurately uncovering the incidence rate for the first year. After that, they may simply assume that the disease prevalence increase between two different years is caused by incidence. In diseases like end-stage renal disease and the treatment of hemodialysis, this may lead to inaccuracies, as a good portion of prevalent patients die each year. Using year-over-year growth as a surrogate for incidence would be highly inaccurate in this case.

Prevalence Population (row 5) may or may not be required, depending on the disease state being modeled. For example, an abdominal aneurysm is most likely a once in a lifetime event. However, in our example of coronary artery disease, patients are likely to have recurring events or medical needs throughout their lifetime associated with the disease. Row 4 represents the percentage of the existing prevalence population that is expected to have a clinical need in any given year. Row 5 is simply taking *Disease Prevalence* (row 1) minus *Disease Incidence* (row 3) multiplied by *Percentage Recurring* (row 4) to derive *Prevalence Population.*

Obtaining the *Number of Diagnosed Patients* (row 6) is simply adding *Incidence Population* (row 3) and *Prevalence Population* (row 5).

The *Diagnosis Rate* is simply the *Number of Diagnosed Patients* (row 6) divided by *Disease Prevalence* (row 1). Diagnosis rate is a valuable calculation as it becomes more difficult to obtain the incidence and recurring population data outside of North America. Many countries do not produce this level of detail and the diagnosis rate calculated from North American or European data can be used as a surrogate to apply to populations that are void of such detail.

Procedural Approaches, row 8, is where this model can become unruly, as you would list all of the procedures and services associated with the disease. For example, you would include all the diagnostic procedures, the medical devices, pharmaceutical products, and long-term care facility or rehab facility costs associated with the disease. These all accumulate to the total cost of this disease state. If you were a health care insurance provider, you would want to calculate all this to determine the total costs associated with this, and all diseases, in any given period of time. Using something like this model in a complex database, health care insurance providers can anticipate costs. Next, the insurance company would add their targeted profit margin on top of all their costs to determine their targeted insurance premiums. The word *targeted* is used as the exercise may result in insurance premiums that are greater than their competitors. Continuing this is beyond the scope of this book except to appreciate that each procedure or service has a reimbursement rate from the insurance payer to the health care provider, such as the hospital or long-term care facility. The reimbursement rate, which is the cost to the insurance company, becomes the revenue line for the health care provider. A health care provider, such as the hospital, would purchase products and services from the various manufacturers. The price the hospital pays the manufacturers is revenue for the manufacturer but cost for the hospital. This is an important concept and you may want to pause and make sure you understand these interactions before you continue. We will come back to discuss these relationships as per Chapter 19 later in this book; for now, let's continue this discussion from the perspective of a life science start-up.

Procedural Approaches (row 8) for coronary artery disease would be medical therapy such as a drug, undergoing a surgical procedure called *coronary artery bypass graft* (CABG), or having an interventional procedure to open up the vessel.

Procedural/Service Approach Percentage (row 9) is the next calculation and this would be derived from research. As you could imagine the health care insurance and providers have very detailed information on these numbers for their populations and they are frequently available to the public.

The next calculation (row 10) would be the *Number of Procedures or Services* used to treat the condition. This percentage is multiplied by the number of *Diagnosed Patients* (row 6) multiplied by percentage of *Procedural/Service Approach* (row 9).

Once the *Number of Procedures or Services* are identified (row 10), you next identify the *Type of Products or Subservices* (row 11) that support the overall procedure or service. We could go on to model the vast number of drugs that could be offered under medical therapy. We could do the same for the surgical procedure of CABG and identify items such as sutures, scalpels, and other tools used in the procedure. However, we will continue using the interventional procedure of stenting that you were introduced to in Chapter 3 earlier in this book. To continue to keep our discussion manageable, we will detail only three of the many products that go into an interventional procedure: stents, a guide catheter, and a guidewire.

Our next calculation is *Units per Procedure or Service* (row 12). This number would be obtained from analytical or research reports. This number can be greater than one for many reasons. In some cases, more than one product could be used during the procedure. In other cases, its extra use could be due to an error, such as selecting the wrong size: once the product is opened, it is no longer sterile and needs to be thrown away. In another case, it could be lost because of hospital inventory management issues. Also, the number does not need to be greater than one. There could be situations where the number is .33, representing that the product is used in only one of three cases.

The *Units per Procedure/Service* (row 12) are multiplied by the *Number of Procedures/Services* (row 10) to derive the *Number of Market Units/Service* (row 13) needed within a timeframe, typically a year.

Next, the *Average Revenue per Event* is entered into row 14. This calculation would again be derived in combination of analyst reports and the company's individual experience.

Multiplying market units by average revenue per unit or event derives the *Market Dollars* (row 15). For example, Boston Scientific Corporation has a significant interventional procedure business. They would take the time to identify all of the products that they sell in row 11. By completing the calculation for all of those products, they would be able to sum them up and determine the potential market for all their business.

The model discussed here would be applicable to the bio pharma industry, the diagnostic industry, and the medical device industry. The biotechnology services industry and the health care information technology (HIT) industries would most likely utilize a different model as their revenue triggers are not patients. This model was used here not only because it is applicable to three major market segments, but also because it allows for two additional illustrations.

THE DISEASE STATE MODEL IDENTIFIES THE TRIGGERS TO RISK AND CREATING MARKET VALUE

Let's return to Figure 11.1 once again and notice the four circled items: *Diagnosis Rate Percentage, Procedural/Service Approach Percentage, Units per Procedure/Service*, and *Average Revenue per Event.* Changing any one of these inputs can dramatically impact the value of the market. For example, a 1% increase in the diagnosis rate in year 2 of the model would add another 252,688 patients to the market. Without changing any other factors in the model, this could add another 31,080 (252,688 × 12.3%) stent procedures. Our last example demonstrated the value that can be created by a small shift in the diagnosis rate percentage. Risk can also be identified using this model. Looking at row 9, the *Procedural/Service Approach Percentage*, one can understand that those who sell interventional products could be concerned that a medical therapy that reverses plaque in heart vessels would reduce the need for an interventional procedure. As interventional procedures have improved over the past 50 years, the number of CABG surgeries has decreased meaningfully.

This model also brings to light the conflicting agendas in health care as manufacturers are each motivated to sell more products and services while insurers are motivated to pay the least amount to keep their populations from needing more health care services.

Let's revisit our earlier discussion around Blue Belt and the venture capitalist's comment, "This is an important technology chasing a relatively unimportant clinical issue." Using Figure 11.1, we can start to understand the comment. The company's original thought was to approach spinal restenosis. If we look at row 10, the *Number of Procedures/Services* would not meaningfully be changed by the use of the company's product. If a second procedure is required today, it is already in the numbers. If second surgeries grew meaningful, certainly the regulatory authorities and hospital administration would question the efficacy of the procedure. One may be able to argue that use of the technology in the first surgery could reduce the need for a second surgery. This would need to be proven in a costly clinical trial and, if it became true, would arguably reduce the number of procedures.

Today, the large amount of bone needed to be removed for an artificial knee or hip inhibits the likelihood of a second artificial knee or hip. These artificial products do not last a lifetime and a bone sparing technology that allowed individuals to get more than one artificial knee or hip in a lifetime would grow the market. Looking at Figure 11.1, row 4, *Percentage Recurring* today is close to zero. The ability to have a second implant would in turn increase *Diagnosis Rate Percentage*, row 7, and materially increase the market's value.

By viewing the components of the *Disease State Fact Book*, we can be amazed that the same bone sparing technology applied to two different rows can either create a yawn or a cheer. Applying the technology to existing procedures does not significantly grow the market's value. Yet when applied to allow for a second implant, the increase in diagnosis rate triggers an increase in the market and inspires venture capital investment.

Using the same model, we can also see risk triggers in our coronary artery disease dialogue. The success of interventional procedures has caused a decrease in CABG surgeries. Interventional procedure manufacturers closely watch the medical therapy segment for fear of a pill that reverses plaque.

THE HEALTH CARE FLOW CHART VISUALIZES THE MARKET'S VALUE CHAIN

Using the *Disease State Fact Book* in our previous discussion helped us identify how to create new categories and increase market value.

The health care systems flow chart (HCSF), Figure 7.1 in Chapter 7 is an overview of the health care system. This tool helps us identify opportunities to collapse or reorganize the value chain in an effort to decrease costs or increase benefits.

Our discussion in Chapter 7 on point-of-care testing (POCT) was a great example of collapsing the value chain. Looking at the HCFC, prior to POCT a physician would take a patient sample and either send it to a hospital laboratory, or send it to Quest or LabCorp, which is represented by the *Distributor* box. With the HemoCue product, physicians no longer needed those components of the value chain and could process the sample in their office to the benefit of the patient.

Looking at Figure 7.1, what other health care technologies or services have you seen evolve in the past few years? Perhaps you have seen urgent care centers appear in your community. If you have been to one, you certainly noticed the convenience of being immediately seen—no appointments are required. You may have also noticed that you can purchase your medication before you leave as opposed to leaving with a prescription and going to a retail pharmacy.

Mobile technologies are also collapsing communication steps in numerous ways. For example, mobile applications are allowing patients to take pictures of their bills and submit them to their health care plan for reimbursement. Historically, one may have had to fill out forms, attach the bills, and fax or mail them to their insurance provider. The provider would have to receive the fax or process the paper. They had to manually validate that the paperwork was in proper order, if not it got sent back to the patient. Today with mobile applications, you immediately receive an error message if you are entering the required data incorrectly. These steps have been significantly collapsed through the use of technology.

Communicating to patients has changed dramatically, as you can create an online account with Quest or LabCorp and see your test results when they are available. Some physicians allow you to book appointments online at their Web sites and fill out your paperwork online before you arrive. Technologies exist to have Health Insurance Portability and Accountability Act (HIPPA) compliant e-mail and text communications between the patient and their medical providers. The next 20 years will be an exciting time, as technology adoption will be accelerated by the need to improve the outcomes of the health care system.

Let's complete this chapter by revisiting the HCSF, Figure 7.1 in Chapter 7. Review one section at a time for familiar companies and recognize that some companies and providers within this chain have more influence than others.

For example, on the manufacturer side, companies such as Johnson & Johnson, GE, Medtronic, Merck, Amgen, and others have significant marketing and sales channel influence. Health care information companies such as McKesson Technology Solutions, Cerner Corporation, Allscripts, and Epic Systems Corporation all have an influential presence.[1] In distribution, companies such as Amerisource-Bergen, Cardinal, McKesson, Owens & Minor, and others have an influential presence. The hospital side has large hospital systems such as HCA Healthcare with 186 hospitals, Ascension with 151 hospitals, CommonSpirit Health with 137 hospitals, and Community Health Systems with 93 hospitals[2] Hospitals are not only influential due to their size in terms of number of hospitals but also based upon their research funding and thought leader influence. Hospitals such as Massachusetts General Hospital, Johns Hopkins Hospital, the Cleveland Clinic, Mayo Clinic, and the University of Pittsburgh Medical Center are all industry influencers. The alternate site providers are defined as those that do not provide in-hospital health care. Surgery centers and physician office practice influencers tend to be regional in nature. Nursing home companies such as Genesis HealthCare with 50,058 beds, HCR ManorCare with 37,079 beds, Golden Living with 29,909 beds, Life Care Centers of America with 26,745 beds, and others clearly influence the marketplace.[3] Top chain drug retailers such as CVS with $109 billion, Walgreens with almost $84.3 billion in revenue, and Rite Aid at $11 billion surely influence access to customers.[4] In addition to drug retailers, durable medical equipment and home medical equipment retailer influences are mostly regional at this time. Companies such as Invacare and Apria Healthcare are starting to nationalize these segments. Large retailers such as Walmart are also entering the space. Returning to the top of Figure 7.1 in Chapter 7, the *Payors* box is represented by companies such as Anthem Inc., UnitedHealth Group, Humana Inc. Health Care Service Corporation, CVS Health Corp, WellCare, and Kaiser Permanente.[5] The Insurance Information Institute reported in 2019 collected almost $964 billion in insurance premiums[6]—clearly influential.

The point of this discussion is to recognize the large number of influencers in the larger health care value chain. Each of these companies has what could be described as channel control over their segment and their customers. Channel control simply means that, based upon revenue size or some other meaningful metric such as beds in the hospital segment or covered lives in the payor segment, these companies strongly influence what occurs in their part of the health care system. It is important to note that these influencers have the ability to move value into their market segments to the benefit of their customers and themselves. However, in doing so, they may not bring benefit to the entire health care ecosystem. This is not to suggest that companies are doing this purposefully. Company investment in market research and value creation tends to be focused upon their specific customer segment and not the entire health care ecosystem.

The importance of this discussion for the start-up is to recognize that health care reform will require that products and services ultimately benefit the entire health care ecosystem: this will be driven by government and private insurance reimbursement methodology changes.

As a life science start-up company can take many years to achieve regulatory approvals and/or an exit, start-ups must anticipate changes in reimbursement methodologies and the potential collapsing of value chains. For those companies that are focused on a specific segment such as medical devices, diagnostics, or pharmaceuticals, they must start to understand how their products impact the entire health care ecosystem. Not anticipating reimbursement's future is the equivalent in the airline industry of not anticipating the rotation of the earth in your navigation formula. In leaving out this crucial piece of information, you will not only miss your destination, you are likely to crash into the ocean. This is clearly a threat for start-ups but also an opportunity. Large corporations are sluggish in adjusting their large infrastructures to such changes. Start-up companies that not only deliver segment value but also health care ecosystem value will be prime acquisition candidates.

For biotechnology service and health care IT companies not committed to the existing methodologies and infrastructure, there is an opportunity to rearrange the value chain components to the benefit of all. The threat is that the larger incumbents have infrastructure and revenue to inhibit the value chains rearrangement. From a start-up's perspective, the challenge is how to plot an exit given this threat. The start-up, once demonstrating its value by customer acceptance and/or revenue, hopes to accelerate an exit, and inflate its price by creating an auctioning environment among the incumbents. To do this, start-ups must create fear that the first one with this ability to rearrange the value chain can shift market shares. The other exit hope for a start-up is that a nonindustry participant sees the product as an opportunity to enter a new market. Given that the non-industry participant is not financially committed to the existing industry's infrastructure, they are only interested in profits and not maintaining the existing infrastructure. Walmart is frequently cited as someone with such capability.

We conclude where we began. Innovation is introducing something new to affect change. To navigate its successful implementation, we must create a model to ensure that we are doing something that is important. The disease state model identifies the triggers to risk and the creation of market value. The HCSF visualizes the market's value chain and also brings to light the conflicting agendas in the health care system. It identifies opportunities to collapse and reorganize to decrease cost and increase benefit. As health care reform will bring simultaneous improvement to quality and a reduction in cost, the health care flow chart brings to light the risk of focusing solely on segment value without considering the benefit to the whole system. Start-ups must anticipate these changes to the best of their ability. Their exit strategies may need to be adjusted as they approach their landing (also known as exit). In the case of the HIT segment, if the incumbents are not ready to acquire, the acquirer may be outside today's market.

DISCUSSION QUESTIONS:

1. Define disease prevalence.
2. Where is disease prevalence derived from?
3. Differentiate "incidence" from "incidence percentage."

4. What is the most theoretically correct way of calculating "incidence"?
5. How is "Prevalence Population" calculated?
6. How is mobile technology collapsing the steps in communication?
7. Explain the purpose and illustrative benefits of the disease state model.
8. What does channel control mean and why is it an important consideration for a startup?

NOTES

1 Top 10 Companies in Healthcare IT Market, 2020, Meticulous Research, October 27, 2020, https://meticulousblog.org/top-10-companies-in-healthcare-it-market/.
2 Replacement: 100 of the largest hospitals and health systems in America 2020, Becker's Hospital Review, Tuesday, December 22, 2020, https://www.beckershospitalreview.com/lists/100-of-the-largest-hospitals-and-health-systems-in-america-2020.html (accessed February 2021).
3 Replacement: Leading Nursing Home Chains in the United States by Number of staffed beds in 2015, Statista, December 2016.
4 Replacement: 15 Largest pharmacies in the US, Becker's Hospital Review, Alia Paavola, March 11, 2020, https://www.beckershospitalreview.com/pharmacy/15-largest-pharmacies-in-the-us.html.
5 Evi Heilbrunn, Top Health Insurance Companies, *U.S. News & World Report*, December 16, 2013, http://health.usnews.com/health-news/health-insurance/articles/2013/12/16/top-health-insurance-companies.
6 New Footnote: Insurance Industry at-a-glance, Insurance Information Institute, 2019, https://www.iii.org/fact-statistic/facts-statistics-industry-overview (accessed February 2021).

12 Demonstrate the Ability to Access the Sales Channel

The previous chapter ended with a discussion on the importance of understanding the triggers that evoke an acquirer's willingness to acquire. This is an important understanding for every start-up company. The start-up must focus on attaining these acquirers' triggers, as without meeting them, an acquisition is improbable. As important as understanding the acquirers' triggers, you need to know what your game plan will be should the triggers not be met or if the triggers are met but acquirers are not ready to buy for internal reasons. For example, an acquirer could be busy dealing with external regulatory matters, a change to internal management, or they may be looking to make another acquisition and your start-up is of lesser importance to them. For example, Boston Scientific Corporation is well known for buying medical device start-up companies. However, when they were working through the multibillion-dollar acquisition of Guidant Inc., all other acquisition activities were temporarily put on hold. Even if other start-ups had met all of Boston Scientific's acquisition triggers, the company was simply too busy pursuing something that would dwarf the opportunity offered by a startup. Start-up CEOs must appreciate this unpredictable fact: when you have met an acquirer's purchase triggers, they still may not acquire the start-up for reasons that are outside of your control.

Ultimately, the only security a start-up company has in this matter is to ensure that the company has the ability to attain revenue and a cash-flow positive state regardless of the acquirer's behavior. To be able to achieve this a company must be able to "materially" enter and access the sales channel. Please note that the word *materially* is in quotes. This is meant to point out that the company must be able to enter the sales channel and obtain a relevant market share over time.

An example may be illustrative. The author mentored a start-up company that had a very interesting product that improved patient safety in a hospital. The product was an incremental improvement to an existing hospital product and was priced at a 20% premium. Appreciate that this 20% premium was on an item priced at a $1 so the premium was 20 cents. Twenty cents does not appear to be huge until you put it in the context that a typical hospital buys 100,000 of these products a year. Thus, this improvement would cost the average hospital another $20,000 a year. Typically, this would result in a "no thank you" from the hospital; however, the company had an answer to this impediment. They could quantify, through an independent study, that the hospital could more than make up for the $20,000 year based upon average annual safety litigation costs. We are moved once again to feel

DOI: 10.1201/9780367533052-12

that this sounds like a viable opportunity. However, our roller coaster ride continues once we realize that this product category exclusively goes through medical surgical distributors. Referring back to Figure 7.1 in Chapter 7, this category of product arrives at the hospital by going through one of three major national distributors. Even though we have the data that may be able to excite the hospital to consider this product, we now realize we must excite the distributors to include this product in their formulary of offering.

Now we have a new constituency that the start-up must demonstrate value to, and to do so, we must understand the distributor's sales model. As the majority of medical surgical products are undifferentiated, medical surgical distributors compete on price. They tend to differentiate themselves through services and other mechanisms beyond product. Given this fact, any price increase would be deemed unattractive to the distributors. Given that most medical surgical distributor's products are commoditized, they are motivated to look for opportunities to compete beyond price. A product focused on safety could be an interesting discussion and provide differentiation. The next question would be if the product category is large enough among all of the other stock-keeping units (SKUs) that the distributor offers to allow the distributor to compete using this particular product. Assuming the distributor wants to continue to investigate, wouldn't the distributor want an exclusive? What would this mean to the start-up's exit strategy? Also one must consider branded manufacturers such as BD, Cardinal, and Johnson & Johnson (called *majors*). They are able to influence the distributor's product offering given their size and ability to negotiate multiple, branded SKUs with the medical surgical distributor. Does the start-up's SKU compete with that of a major? If so, what are the implications to the major/distributor relationship? Is it worth the distributor taking this issue on?

All of the above speaks to the author's frequent statement that "technology de-risks failure but not an exit." You can appreciate from the story that a company could have a perfectly performing product with demonstrated benefit but is inhibited by its ability to access the sales channel. This is why the start-up must understand their ability to gain sales channel access before they raise their first dollar of funding.

Without identifying the specific company, the end of the story is that the young start-up CEO was terminated and a more seasoned CEO was hired. The first thing that the new CEO did was to recognize the improbability of going through distributors and he created a new strategy to access the market. Time will tell if this new strategy will work but today they have revenue and are on their way to cash-flow positive.

Sometimes a start-up has a different challenge. In another example, a biotechnology tool start-up had the ability to access the sales channel but the geographic distribution of the customer made it too costly for them to distribute. At the time they formed, the start-up was aware that existing manufacturers had a similar product. They falsely assumed that the manufacturers would be excited to have a better performing product. When this situation was realized, the start-up decided they could compete directly with the national manufacturers and once demonstrating sales, one of the distributors would acquire them. The company decided that they would sell the product directly to their customer themselves as evidenced by their success at trade shows. There were three major trade shows a year that this company went to and after each trade show, the start-up had a huge list of interested customers—called *leads*. Subsequently, they also learned that they had an unusually high success or hit rate with their trade show leads, frequently converting them into a sale. So what was the problem?

The problem was that the industry required the company to pay an average of $80,000 a year for an individual sales professional. If you assumed that the company could afford the sales expenses of 16%, each sales rep would have to sell $500,000 a year to make the company financials be aligned with industry standards.

Let's pause and make sure we are calibrated on what "aligned with industry standards" means. An income statement is used as a scorecard of annual profitability. An income statement starts with revenue from the sale of the product from which the cost of building the product is deducted. The cost of building the product is called *cost of goods sold*. Revenue minus cost of goods sold is called *gross margin*. After gross margin, the company deducts those expenses that are not tied directly to the production of the product such as research and development, finance, marketing, and the sales departments. These expenses are generally referred to as SG&A, which stands for sales, general, and administrative expenses. After gross margin SG&A is deducted to arrive at net income before taxes and adjustments.

When an acquirer looks at buying a start-up company, they want to make sure that the start-up's gross margins are aligned with the acquirer's gross margins. In situations where the acquirer will not use the start-up's SG&A they may not be concerned, because the acquirer might be putting the start-up product through an existing sales force—costing them very little incremental cost to sell the start-up's product. In cases where the acquirer is interested in using the start-up's existing SG&A, alignment is important.

Returning from our pause, the biotech tool start-up company needed to create a business model that allowed them to prosper while waiting for an acquisition, which was not plausible in the short term. For the company to get acquired, they were going to have to become large enough and differentiated enough that the acquiring distributor would be willing to terminate, sometimes called *sunsetting*, their existing revenue line in the hope that the new revenue line (the start-up's) would result in a more profitable future than if they did nothing. It would take some time for the start-up to prove this to an acquirer. This returns us to the start-up's immediate problem of paying a sales professional $80,000 in the hope that he/she could generate revenue of $500,000. The question is, would this work?

Another pause is required to discuss the process of a sales call. Fundamentally, every sales process is no different than any other process such as manufacturing a product. In manufacturing a product, one would create a bill of materials consisting of all the components that go into the process. The labor steps to create the product would be detailed and the appropriate materials and subassemblies would be added to each labor step to culminate in a finished product. It would be recognized that the manufacturing process itself would not be perfect and be subject to losses along the way. This is referred to as *yield loss*. Yield loss costs both time and money that would be incorporated not only in the cost of the product itself, but also in calculating the capacity of production.

A sales process is no different. The sales process starts with understanding how many sales locations there are. Next, one would determine how many target opportunities exist in a particular sales location. This would provide us with the sales opportunity for the location. Next, the company would determine what is called a *hit rate*. Hit rate is the assumption that when working with the targeted customer that a sale would ultimately occur. For example, if a sales professional goes to 100 accounts and 50 accounts are interested in buying, the company would have a potential hit rate of 50%. After determining hit rate, the next question is how many visits are required and the conversion time before the start-up would expect to receive its first sale. For example, does it take four sales professional visits over the course of 6 months to make one sale? This is important to know. After the start-up receives its first sale, what is its reorder rate? That is to say, how long will it be before the customer places another order? For example, if we were discussing bottled water, one perhaps would buy weekly. If I were buying a new car, perhaps the reorder rate is every five years. After one determines the reorder rate, the next question is what is my retention rate? Returning to our bottled water example, if data shows that once a customer buys a particular brand of water and they repurchase (reorder) it weekly for 52 weeks, one could assume in a sales forecast that this individual would continue to buy the product.

Let's take a look at the table in Figure 12.1 to illustrate how that biotech tool company appeared attractive on the surface but then a thorough examination of the sales process became unattractive. The table shows that there are 36,000 potential targets annually for this product. At an average selling price of $650, this generates an annual market size of $23.4 million. On the surface this looks very exciting, however, let's take a moment and walk through the details of Figure 12.1.

The table shows a fundamental problem. The company loses money with every sales professional that they hire. How is this so? This appears to be a simple problem of geographic reach and capacity. There are only three sales targets per location and the sales professional takes over 600 minutes to convert one client. The other issue identified in this model but not placed into its calculation is that it takes nine weeks to convert one client. Detailing this model further is beyond the scope of this book, however, the reader should be aware that the model would need to create a capacity constraint for each individual sales professional as it takes nine weeks to convert an account and there are only so many product sales hours in a day, which is limited by geographic reach. Continuing with the model above, cost of goods sold equals 55%, allowing 45% of the sales price to be available to assist the company in profitability. The fundamental problem with this company was that the geographic dispersion combined with the low number of customers per account perpetually inhibited their ability to profitably capture any share of the $23.4 million market.

This example demonstrates a frequent miscalculation a start-up can make. By solely focusing on the size of the market and projecting the share of market that the start-up feels they should be able to take, the company erroneously projects success. For start-up companies to be successful, they should reconcile their market projections by performing a similar analysis to the above. If the market share calculations calibrate to the sales analysis, then the company can comfortably proceed. If, as in this case, they do not reconcile, the company must either change their

	Annually	Unit Price	Market Size
Sales locations	12,000		
Targets per location	3		
Potential targets	36,000	650	$ 23,400,000
Hit rate per location	75%		
Potential hits	27,000		
Visits to conversion	4		
Average conversation time	9 weeks		
Average drive minutes between visits	60		
Average minutes per customer visit	90		
Time of one visit (drive + visit)	150		
Total minutes to conversion (4 visits)	600		
No. of annual sales minutes per rep	120,000		
Potential conversions per sales rep	200		
Average sales price per conversion	650		
Revenue potential per sales rep	130,000		
Cost per sales rep	80,000		
Cost of product 55% of sales price	58,500		
Contribution per sales rep	50,000		
Loss per sales rep	–28,500		

FIGURE 12.1 Cost of customer acquisition.

approach or not proceed. In this case, the company decided to proceed and within a year went into bankruptcy.

In the end, both of these examples demonstrated the need to understand the ability to access the sales channel to derisk an exit. In the first example, meeting and acquiring the start-up's acquisition triggers was interrupted by an event that the start-up could not control—a lawsuit. As a result, the start-up must be able to survive beyond its predicted acquisition trigger. In the second example, an attractive $23.4 million market clouded the realities that the sales model selected could not achieve profitability. This was compounded by the fact that the potential acquirer already sold what was perceived by them as a similar product line. To change the acquirer's mind in this situation would require a meaningful demonstration of revenue. Understanding the situation and realizing the sales model that was represented, the biotech start-up should have never gone forward.

This chapter speaks to one of the biggest weaknesses that the author sees in the formation of life sciences start-up companies. As scientists frequently initiate life sciences start-ups, the company's first leaders tend to have limited sales experience. As a result, they skip over the important analysis of determining access to the sales channel. This frequently results in the start-up failing after it has raised capital; thus, it is irresponsible not to conduct such an analysis before accepting investor capital.

DISCUSSION QUESTIONS:

1. What is meant by a company to “materially” enter and access the sales channel?
2. Discuss the concept of de-risking from the different perspectives of the customer, the investor, and the acquirer?
3. What is the cost of customer acquisition and important consideration?
4. What is “yield loss”?
5. From the values in Figure 12.1, what was the fundamental problem with the company?

13 Gather Domain-Experienced Personnel to Reduce Risk

Hungarian author and playwright Frigyes Karinthy introduced the concept of six degrees of separation in his 1929 short story *Chains*.[1] The concept is that everyone and everything is six or fewer steps away from another person in the world via an introduction from any other person. The implications of this are important to the start-up. Acquirers like to acquire from people they trust and funders like to fund people they know. As the start-up considers the people they are going to involve in their company, they should consider the relationships of these individuals to the customer, the acquirers, and the funders.

Malcolm Gladwell, in his book *The Tipping Point: How Little Things Can Make a Big Difference*, speaks of the magic moment when an idea crosses a threshold causing it to tip and spread like wildfire. One of his examples in the book is how a single sick individual can ultimately cause a flu epidemic. This demonstrates the power of "patient zero"—the one person who started it all. The tipping point explains how there are three kinds of people: Connectors, Mavens, and Salesmen. Connectors are those individuals with a big contact list: their contacts tend to be broad. Mavens are the trusted experts. Salesmen are those articulate individuals who can pursue others and push new concepts into adoption.

For the start-up, finding domain-experienced individuals seeds your Connector network, as these people have significant associations within your particular industry (domain). Connectors can bring together the team of people necessary to get the job done. Mavens are trusted for their expertise and are looked to validate the start-up's technology or business model. The Salesmen are critical in taking the Mavens' expertise and the Connectors' network and accelerating the start-up's objectives.

Connectors are the ones who most likely have relationships with funders, customers, and acquirers. It is rare for someone to have all three types of relationships—generally, they have two. The most general groupings are Connectors who have funder and acquirer relationships or Connectors with customer and acquirer relationships. Connectors who operate at the macro level of the market are more associated with funders and acquirers. Connectors who operate at the micro level (product level) of the market are more associated with customers and acquirers.

For example, a Connector physician who works daily in the acquirer's and start-up company's target market works at the micro level. His expertise is in products and his knowledge of service areas such as general surgery, cancer treatment, rehabilitative care, and so on. Since the majority of start-up acquisitions are by companies already selling into a specific segment, acquirers most likely know these

DOI: 10.1201/9780367533052-13

Connectors as they already sell products to them. These Connectors, both known and trusted, are frequently called *thought leaders*. These thought leaders are very valuable to the start-up company as they not only provide great guidance on product specifications to the company but also frequently provide feedback to potential acquirers. Some of the more famous thought leaders are also known to corporate and venture capitalists. When known, they are frequently consulted during due diligence to determine if an investment is advisable. These funders look to these physicians to validate that the product is solving an important customer problem.

Moving from the micro environment to the macro environment, we find funders and acquirers. A funder's goal (venture fund) is to raise money, invest in a handful of companies, and exit within a certain time period. Venture capitalists will then raise another fund and do it again. In larger firms, there can be more than one fund existing at the same time. To facilitate exits, the venture capitalists must have relationships and a favorable reputation with acquirers. Acquirers may also have a corporate venture fund. Corporate venture funds frequently invest alongside venture capitalists and venture capitalists frequently invest alongside other venture capitalists. What this means is that the funder not only has a connection with acquirers but also has connections to other funders. Funders with a strong reputation for uncovering valuable start-ups will be followed by acquirers as well as other funders. Acquirers and other funders recognize the reputation of the funder's ability to select good companies.

Returning to the book *The Tipping Point*, one of its best examples of the value of the Connector is the discussion of Paul Revere and William Dawes during the British Army's arrival in 1776. Both individuals rode into the countryside at the same time warning of the British advancement. As a Connector and a Salesman, Paul Revere had better results communicating the warning and sparking action. Revere's warnings were taken more seriously, as he was known and trusted in numerous social circles. As a result, people did not question his motives and his information was immediately trusted. William Dawes was not as effective, simply because he was not known. As he rode past individuals, he was most likely either ignored or his information required affirmation before it was believed and passed on. This was not a constraint for Revere.[2] In a start-up company, time is money, as evidenced by the concept of burn rate. If burn rate is the amount that is spent per month, each month that it takes to get acquired requires additional money. Paul Revere's story exemplifies that a trusted Connector can quickly reduce the time it takes to get through a network and achieve a result.

Let's move from Connectors to Mavens. Mavens are the experts and their expertise perhaps may not be in what one would normally associate with the word. Upon first thought, one would think of these individuals as the academics or scientists that are experts in their particular field of study. This would be true but it would limit the definition of a Maven. The Maven may have expertise in finance, quality, regulatory, or operations. For example, we previously discussed in Chapter 13 the physician as a Connector but he can also be a Maven: the expert, the knowledge leader in his area of expertise. He could be the thought leader, oncologist, cardiologist, or neurologist in his field. Every physician specialty has a handful of these individuals. For the start-up company, a positive opinion about the start-up from a Maven is a valuable validation point. This validation point may also lead to a fundable milestone for the start-up company. Previously, we discussed in Chapter 4 the value of an SBIR in validating one's technology. This occurs because an SBIR grant is scored by a committee of Mavens, thus providing the independent assessment that offers due diligence confidence to an investor or acquirer. Another example of a Maven could be in the area of regulatory affairs. In the life sciences segments (verticals) of pharmaceutical, medical devices, and diagnostics, approval from the FDA is required. Frequently, there are multiple ways in which a regulatory path could be selected and these pathways equate to risk and cost for the start-up. If a Maven in the regulatory field bestows an opinion on the pathway in the clinical endpoints, this can provide clarity for the start-up and independent validation and a de-risking point for funders. These examples should help you broaden your initial perception of a Maven and recognize that Mavens exist throughout the start-up's social network.

The expertise of the Maven cannot be realized without the Salesman. The Salesman is not just the functional position of sales. They are charismatic people who are skilled in the art of persuasion and have the ability to negotiate. They are adept at learning from their Mavens and have the ability to connect with people and express the Maven's vision. They are skilled negotiators and have the ability to handle objections and questions. Every question a salesman receives that he cannot answer is deemed a learning experience as he goes back and tries to determine how to answer the question in the future. His motivation is to continuously improve his ability to answer the questions or objections in such a manner that the receiver can process the information. This is an important trait of the Salesman; his desire to influence demands that he answer questions in a clear way. It is important to note that a Salesman can also have traits that overlap with the traits of a Connector or Maven. For example, a successful venture capitalist can persuade other corporate and venture capitalists to invest in a particular firm. Many Maven scientists and physicians also possess the ability of the Salesman. Regulatory affairs Mavens have the ability to listen to the FDA's concerns and seek to translate company information into a form that satisfies FDA's concerns. A patent attorney Maven may be skilled in working with the patent office. He may have a charismatic capability to form a relationship with the patent officer and create a two-way dialogue that leads to the creation of a patent that satisfies both parties. What is common among all these Mavens and Salesmen is the recognition that it is necessary to possess the Salesman trait to advance vision into adoption. Thus, the Salesman has the ability to accelerate the speed of the creation and expansion of the start-up's social network.

Before we conclude our discussion on these archetypes, it is important to note again that although there are three kinds of people, Connectors, Salesmen, and Mavens, all are likely to possess some combination of these traits. Remember the example of Paul Revere: he was both a Connector and a Salesman. One could argue that he had Maven traits as well, as his skill as a silversmith was well noted in history.

As the start-up ponders the social network it needs for a successful liquidity event, it is important to select the team based upon a number of factors. One of the criteria that should be added to the mix is determining the appropriate combination of Connectors, Salesmen, or Mavens in the creation of the start-up's network. For example, what if William Dawes was aware of this information before he set off on his ride to warn of the British troops' impending arrival? What if he knew that he did not possess the skill of a Connector or Salesman, what could he have done? Frequently, the founder of a life sciences start-up company is a Maven, given that the industry is founded on science. Is not the scientific founder facing the same challenge that Dawes faced? How could he, or any start-up for that matter, create a model to help them solve this problem?

Let's pause for a moment and revisit the themes and concepts regarding due diligence that were discussed in Chapters 8 and 9. In Chapter 8, we discussed that a start-up requires a continuous stream of due diligences. Each subsequent due diligence builds upon the success of its predecessor, leading to a reputation and a resulting liquidity event. In Chapter 9, we expanded our discussion and expressed that due diligence is a critical brand component and requires a supporting business process. Without a supporting business process, the company increases its risk and faces reputational harm. We concluded Chapter 9 by referring to a "Due Diligence Checklist" that can be found on the book's Web site, at: https://healthcaredata.center/commercialization-startups that summarizes a supporting business management process.

How does a discussion on due diligence relate to this chapter's theme that domain-experienced personnel reduce risk? If we turned our "Due Diligence Checklist" into a graphic, it would look like Figure 8.1 in Chapter 8.

Take a moment and really look at Figure 8.1. Doesn't the diagram appear to be a network diagram? Doesn't the due diligence process summarize into a network diagram? Doesn't this network diagram express the social network of a start-up? Isn't the goal of this diagram to ensure that the start-up is gathering all the appropriate Connectors, Salesmen, and Mavens to validate the reputation of the company?

If we accept that our business process that supports due diligence leads to the start-up's network diagram, can we also recognize that the start-up's network diagram is supporting its marketing brand? We will discuss this more later, but it is important to recognize that as you add people and organizations to your network, you are indeed making a branding decision.

Let's walk through this network diagram and discuss the value of each component.

GOVERNMENT

It would be impossible to create a concise graphic that includes all of the components of how government influences or provides domain value to the start-up company. However, a summary of some of the more important features is offered for your consideration.

Types of Legal Entities

Although today a company can easily be formed with the assistance of the Internet at a very low cost, it is advisable to consult an attorney to ensure that you are building your business on a solid legal foundation. There are four basic types of structures that a company can take: a corporation, a general and limited partnership, a LLC, or a sole proprietorship.

Before we discuss the various types of legal entities, we must first understand the start-up company's objective. As the author is associated with an incubator that has both economic development and for-profit funds, he sees two kinds of start-ups.

The most prominent start-up type is that in which the company intends to raise capital and create a liquidity event via an acquisition, a merger, or an IPO within a 10-year time period. This goal aligns with corporate and venture capitalists' funding objectives and a subset of angel investors. These companies generally enter emerging and large markets and anticipate rapid sales growth, which precipitates a liquidity event. This book is primarily written toward this type of start-up, as this model begins with the intention of creating something that is temporary and requires a liquidity event for its stockholders. For clarity in our upcoming discussions, we will call this type of start-up a "Type 1" start-up.

The second type of start-up company is one that enters a smaller market or a mature market and hopes to build revenue and provide profit over time. Although they may eventually get acquired, they are equally happy to generate and distribute profits along the way. These types of start-ups are more attractive to angel investors and are not generally of interest to corporate or venture capital. For clarity in our upcoming discussions, we will call this type of start-up a "Type 2" start-up.

Given the above, let's discuss the types of legal entities applicable to businesses and start with the sole proprietorship. A *sole proprietorship* is generated by one individual for his own benefit. It is not considered a separate legal entity from its owner, although it is separate for accounting purposes. The proprietor risks his personal assets in satisfying the liabilities of the entity. This is clearly not an appropriate structure for the start-up companies that we speak of in this book.[3]

A *general or limited partnership* (LP) is an agreement between two or more persons who join together to carry out a venture for profit. Partners contribute money, property, and/or labor, and each has unlimited personal liability for the debts of the business, although the liabilities are capped according to the amount of money invested by the individual. Although not appropriate for a start-up company, this may sound somewhat familiar to the reader. The reason for the familiarity is that general or limited partnerships are utilized by angel networks and venture capital organizations. These structures work for those organizations as they are putting money to work via an investment in the hope of obtaining a return. The losses that they risk are in the amounts that they invest. Please refer back to Chapter 6 to reference our previous discussions on this topic.[4]

A *limited liability corporation* (LLC) has aspects of a general or limited partnership and a corporation. Participants in the LLC are called *members* and have limited liability like the stockholders of a corporation. There are no limits on the number of members. Like an LP, LLCs have the tax advantages of passed-through taxable income or losses. In corporations, which we will discuss momentarily, profits are taxed at the corporation level and then again at the individual level when a dividend is received; many refer to this as the double-taxation effect of a corporation. LLCs do not have to observe such formalities imposed on corporations as annual reports, stockholder meetings, and other shareholder requirements. These activities cost money and these expenses can be avoided in an LLC. Unlike corporations, which must distribute profits in proportion to each shareholder's ownership, LLCs are able to distribute profits in any manner they choose. An LLC is applicable to both Type 1 and Type 2 start-ups. It is highly probable that a Type 1 start-up will eventually need to convert to a corporation to satisfy the needs of corporate and venture capitalists. For the Type 2 start-up, an LLC is the most effective mechanism for distributing profits, as the company may or may not be planning to be acquired.[5]

A *corporation* is a legal entity with a defined scope of activity and a charter. The company's charter restricts the company's activities. For example, a company with the life sciences charter cannot all of a sudden shift its focus to energy. Additionally, there are higher reporting requirements. What is attractive about the structure is that stockholders are completely protected from liability, although this comes at the cost of double taxation. So why would a Type 1 start-up desire to be a corporation over an LLC?

Corporations are designed for perpetual life, resulting in an easier ability to transfer ownership. Although it is most likely obvious to the reader, it should be noted that the ease of transfer differs between a publicly traded stock and a private stock. A publicly traded stock can easily sell on a public exchange and is highly liquid. A private company's stock may have restrictions, such as in the case of a stock offering giving the company the first right to buy the stock. Unlike a publicly traded stock that has an exchange to communicate the wants and needs of buyers and sellers, to trade a private company stock its owner must find a buyer. For those who participate in a Type 1 start-up, this risk is offset by the upside potential of the future liquidity event. Another attractive aspect of a corporation is that it is easier to raise capital to pursue growth by simply issuing more stock for sale. An LLC, for example, would require the rewriting of all the member agreements. Appreciating that a Type 1 start-up is designed for an acquisition, merger, or IPO, one can see why start-ups prefer this structure.

An *S Corporation* might be the appropriate place to begin to work out one's structure. Although this election limits the number of members that can participate in the S Corporation, this may not be an issue in the short term and allows for the avoidance of double taxation as profits and losses are reported by the owners on their individual tax returns. This may appear to the reader to have the benefits of an LLC and on the surface this is true. However, there are stricter operational requirements of a corporation, such as director and shareholder meetings, governance bylaws, and annual reporting requirements. Additionally, the shareholder of the S Corporation must receive "reasonable compensation." The intention of this requirement by the IRS is to ensure that the corporation pays appropriate payroll taxes.[6]

Let's return to Figure 8.1 in Chapter 8 for a moment as we reflect upon why we need to ponder the long-term goal in forming the company when we select the type of legal entity. Can't the start-up simply pick the cheapest one to get going and convert it later?

If we recognize, as we do in Figure 8.1, that every activity builds to create a due diligence network and supports the brand message, we recognize the importance of this consideration. From a due diligence perspective, selection of a legal entity requires a supporting business process. In the spirit of efficiency and risk, it makes sense to do it once and have continuity in its supporting business process. From a brand perspective, the start-up's selection of an entity can speak of the founder's intent. A story may be illustrative.

As previously stated the author is associated with an incubator. Several years back, at a networking event, the author purposely connected the founder of the start-up with a high-net-worth angel. After the introduction, the author left the two alone to have a conversation. Later in the evening, the author found his way back into a conversation with the angel. The author asked the angel if he was interested in investing in the company and was surprised that the angel said no, he was not interested. When the author asked the angel if his concern was the start-up's technology, the angel commented that his concern was that the company was a sole proprietorship, which told him that the founder really did not want to pursue the start-up model. The founder's choice told him that the founder wanted to be in control and be the boss more than he wanted to be rich. The angel was only interested in investing with individuals who were more interested in generating large returns.

This speaks to the adage that first impressions are difficult to change. In fact, evidence suggests that this angel would only change his mind about the company should this founder no longer be its leader. Dr. Rick Nauert started his career as a clinical physical therapist. In an article published in *Livescience* titled, "Why First impressions Are Difficult to Change," he speaks to the evidence of the context of a first impression. In this article, he created a study that provided participants with either positive or negative information about an unknown individual. Later in the study, the participants received new information about that same individual that was inconsistent with their first impression. The receivers of the new information contextualized the new information so that they could reconcile it with their first impression. For example, if the initial impression was that an individual was a bad business person, the new information could cause the receiver to recognize that the person was perhaps a great father. The new information did not, however, lead the receiver to change his mind about the preformed opinion regarding the person's business sense. As the study continued, it was discovered that in order for the first impression to lose its power, it had to be challenged in multiple different contexts. As one can imagine, for a start-up company it is not only difficult to have multiple opportunities to change the context of an impression, it is also time-consuming and expensive.[7]

The author shares a personal story that exemplifies this point. The author joined a division of a Fortune 50 company in the month of September. Unfortunately for the author, his predecessor knew he was leaving the company and did no preparation for the 5-year strategic plan that was to be presented to the corporation's chief operating officer within 9 weeks of the author's joining the company. The author worked almost 80-hour weeks to prepare for this presentation. The author's boss, the division's president, was empathetic to his difficult situation and had communicated the challenge to the chief operating officer of the corporation. The president was kind enough to come back to the author and state that he did not need to "kill himself" for this presentation. However, what the author knew is that he only got to present to this chief operating officer once a year. If his first presentation were bad or neutral, it would take another year for the author to present again and provide the chief operating officer with a favorable opinion. 2 years out, the author knew that the chief operating officer would have one good opinion and one bad opinion, and it could take several years for the chief operating officer to have more favorable experiences than not. For the author, he recognized that that first impression could cost him several years of work and thus the short-term extra level of effort actually was the least amount of work in the long run for the best chance at a good result.

Both the author's story and the angel's story demonstrate that every start-up interaction presents an impression. Figure 8.1 in Chapter 8 demonstrates the vast number of participants required to deliver a favorable, branded impression. Although at this point we are simply speaking to the decision of electing your legal entity, the stories above speak to what appears to be a relatively small decision communicating much about the company.

Patents and Trademarks

Having a patent gives you the right to exclude others from replicating your approach or product. Having a trademark establishes your exclusive right to use the trademark in connection with goods or services.

Looking at the above, one can appreciate the blocking power created by having both the patent and trademark. Maximizing that value is best done by individuals who have deep domain experience.

In the case of a patent, the domain experience may be provided by an academic, a scientist, or someone within the R&D or manufacturing departments. A patent can be achieved by communicating uniqueness in what the product does or how it is manufactured.

For example, in medical devices it is not uncommon for a physician or an R&D engineer to be the initiator of the patent. As the physician is the expert in achieving clinical results, he is also the expert in the use of the tools that create those results. The desire to constantly improve clinical results inspires creativity to create new tools. The R&D engineer can also be the source of technical expertise as these individuals are constantly exposed to physicians and sales and marketing specialists. The R&D engineers collect information from the people with whom they work, which unlocks their creativity and enables them to use it to solve problems. Individuals in manufacturing can also use their experience to improve outcomes. They can create new methods or improve existing methods.

Trademark domain experience generally comes from those who focus on commercial activities such as marketing, sales, or public relations. The objective of creating a trademark requires an understanding of the hierarchy of thoughts that a customer processes when selecting a product or service. Understanding this hierarchy requires expertise and domain experience. If the trademark can capture a word or an image that comes to the top of the hierarchy, the trademark can generally garner more sales than the product could achieve without the mark.

There is no better representation of this than Coca-Cola. When John Pemberton created the formula for his new drink in 1886, he sought to create a trademark that distinguished his product and there is no doubt of his success.[8] The goal of every trademark is to have it define a category. For example, have you ever gone into a restaurant and asked for a Coke and had your waiter or waitress say, "We don't serve Coke, is Pepsi okay?" How many advertising dollars did Pepsi have to spend to get the waiter or waitress to say the word Pepsi? In an article published in 2012, the author Mallory Russell commented that between 2010 and 2011 PepsiCo significantly increased their advertising and marketing spending and achieved a 38% higher sales growth year-to-year than Coca-Cola did. However, she pointed out that with all that effort, PepsiCo only sold $12 billion while Coca-Cola sold $28 billion. This speaks to the potential power of a trademark.[9]

Grants

We have previously discussed in Chapter 4 the value of government grants and demonstrating that achieving one offers an independent third-party validation of your technology. No further reiteration of this value is needed here, except to point out that the value of the validation resides in both the reputation of the grant authority itself and the domain expertise of the individual evaluators on the grant committee.

The start-up must ask itself how it can form an appropriate relationship with the individual grant authorities and the individual evaluators. For example, many life science start-ups have founders from academia. These founders most likely have received academic grants in the past. All of these academic submissions bestow a positive reputation upon the academic. If an SBIR grant is submitted and a known academic is a participant, the reputation of the academic favorably contributes to the outcome. Ironically, the academic has built his reputation just like Coca-Cola has, albeit in a different manner. The academic's history of solid research is a surrogate trademark. And that trademark avails its value when the review committee sees the academic's name on a proposal. If two proposals were of equal merit in every way but one possessed a "trademark" academic on it, it is more likely to get funded than the proposal that has unknown participants. Thus, a pre-seed and seed stage start-up company should work as hard to know these individuals as they do to know their ultimate customer.

Regulators

Obtaining regulatory approvals is one of the most costly activities associated with a life sciences start-up. An article posted to the Medical Device and Diagnostic Industry Trade association's Web site (MD&DI), stated that in 2019 the average cost of obtaining 510k approval was $31 million and the PMA pathway was $94 million in the United States.[10] It should be noted that this class of FDA approval is considered one of the least expensive regulatory pathways. Our discussion here is not to document all the regulatory standards that a start-up must meet but to offer a sample.

The FDA

The FDA Web site states that they are responsible to ensure public health by ensuring safety, efficacy, and security of human and veterinary drugs, biological products, medical devices, food, cosmetic, and other products that admit radiation.

FDA regulations are constantly evolving and an individual with deep regulatory experience in your start-up's vertical and historical relationships with the FDA is invaluable.

A start-up's regulatory pathway is one of the most, if not the most, expensive aspects of commercializing a life sciences technology. Making a mistake in the appropriate selection of the FDA pathway, and/or in the execution of the clinical trial, and/or in the submission of the data can literally bankrupt the company. Start-up funders recognize this risk and are hesitant to make an investment until this risk is reduced. For later-stage investors, the risk is only reduced by FDA approval, for early-stage investors that risk is reduced by the FDA concurring with the clinical trial pathway. For pre-seed and seed investors, their assurance comes from either an independent opinion from an external law organization or by hiring a domain-experienced regulatory professional. It should be highlighted that the pre-seed and seed investors are relying upon the domain experience of the domain-experienced regulatory professional at the external law organization to reduce their risk.

Unfortunately for many start-up organizations, they do not tie or align risk reduction techniques with the financing stage of the start-up company. The start-up must recognize the importance of this to their financing in the perceived risk profile of their company. The story on CLIA labs in the following section also demonstrates this point.

Centers for Medicare & Medicaid Services

The Centers for Medicare & Medicaid Services (CMS) mission states that they foster health care transformation by finding new ways to pay for and deliver care that can lower costs and improve care. This aligns with our previous formula for outcomes:

$$\text{Outcomes} = \uparrow \text{ availability} + \downarrow \text{ costs} + \uparrow \text{ quality}$$

Revisiting Figure 7.1 in Chapter 7, our health care systems flow chart visualizes that CMS's reimbursement hierarchy is the backbone of the nation's (U.S.) health care system. Even if private payers do not utilize CMS structures, private payers' structures are philosophically aligned to CMS's payment rationale. It is important to note that each country has its own reimbursement processes that must be known for the start-up to prosper. However, we will continue to use the U.S. reimbursement system as an example as it is more highly documented than other nations. Documentation should not be confused with simplification: it is still an incredibly complex system.

The ability to maximize the reimbursement for one's product is of the utmost criticality to a start-up company. Funders are going to need confidence that a company's reimbursement strategy is not based on naïveté. Additionally, a start-up company must have enough foresight to realize, appreciate, and align to an acquirer's reimbursement strategies. During due diligence, an acquirer could discover your reimbursement strategy contradicts theirs; this could be a major obstacle to your acquisition.

A start-up company must demonstrate prudent removal of this risk to funders. The exact demonstration of this risk removal will be different depending upon the stage of the company and the particular vertical such as pharmaceutical, medical devices, diagnostics, and so on. Involving domain-experienced personnel early is critical to one's success.

One of the biggest opportunities for synergy for a start-up company is to recognize that in the execution of their FDA requirements, there may be an opportunity to simultaneously collect data necessary to achieve CMS reimbursement. This is frequently overlooked and adds time and cost (burn rate) to the organization.

One vertical that is particularly challenged in this space is the diagnostics category, specifically molecular diagnostics. The FDA standard of CLIA (Clinical Laboratory Improvement Amendments) establishes quality standards in laboratory testing to ensure accuracy, reliability, and timeliness of results. Passing CLIA standards does not demonstrate that the product results in something that is clinically important. As molecular diagnostics requires significant scientific expertise, their founders frequently come from a scientific background. As obtaining CLIA is not as expensive as a medical device or pharmaceutical FDA milestone, angels tend to gravitate toward this category without appreciating that these companies need to spend tens of millions of dollars more to obtain market studies to both demonstrate clinical importance and reimbursement rationale after CLIA certification is received. The scientific founders, lacking this domain experience and understanding, frequently raise only enough capital to obtain CLIA certification but not enough to demonstrate clinical importance or convince CMS to reimburse their test. When this occurs, funders lose money and founding management gets replaced.

CE Mark and Foreign Regulators

CE Mark is an example of a European Union medical device requirement. The CE Mark is a legal designation that a medical device product meets all relevant requirements in the European Union's medical device directives.[11] Medical devices cannot be sold in Europe without this mark. This is only one example of the many types of foreign regulations that a life sciences company faces: there are too many of them to include in Figure 8.1 in Chapter 8.

The importance of this discussion is to recognize, as with FDA and CMS regulations, that the activities associated with entering a foreign market are complex. These activities are costly and a delay due to poor execution not only results in increased burn rate, but also delays the start of revenue. Thus, obtaining domain expertise is absolutely critical in preparing to enter foreign markets. Selecting regulatory personnel that has experience in foreign registrations is often not enough. Even with domain experience, it is highly unlikely that the start-up could ever afford to bring sufficient expertise in-house. In an effort to apply domain-specific resources and demonstrate confidence to those performing due diligence, the start-up company must add branded channel partners to assist them in preparing these registrations.

ISO, cGMP, and Other Process Certifications

Most life sciences start-ups are required to meet various internal process standards. The Internal Standards Organization (ISO) 9000 family is a group of quality standards designed to ensure that there is a supporting management system to deliver products and services to meet customer expectations. These standards are published by ISO. It is important to note that ISO does not perform the certification and the company must choose a certification body. One of the biggest mistakes start-up companies make is in selecting the most inexpensive certification body; this can frequently result in a more costly and time-consuming process.[12]

cGMP stands for Current Good Manufacturing Practice regulations. These regulations are set by the FDA for drugs and medical devices to provide the minimum requirements for the methods, facilities, and controls used in manufacturing, processing, and packaging of drugs and medical devices.[13] These regulations incorporate design controls to ensure the proper development of a product and operational controls to ensure that appropriately developed products are produced consistently and the supporting processes such as manufacturing, quality, warehousing, and others stay in control.

Life sciences start-ups are frequently founded by individuals with a background in science, research, and development, business development, marketing, or sales. These individuals often have no experience with these regulations. Some start-up companies obtain these in-house resources after the seed financing phase. Without appropriate guidance, many companies must pause and regroup to prepare for the regulatory submissions. Frequently during preparation meetings, the company may find that a technical brick in the foundation of this submission is either missing or found to be erroneous. The cost of the delay can be devastating to the start-up company. If one appreciates that a submission to the FDA results in a value milestone, and frequently a fundable milestone, a stall at this phase could cause the company to run out of money or have a down round (devaluation).

CHANNEL PARTNERS

Value Chain

Let's return to Figure 7.1 in Chapter 7, the health care systems flow chart, to visualize the number of channel (value chain) partners that could be involved in delivering your product. Channel partners allow your product to get into the hands of your customer. They can also offer a supporting component that contributes to your customer's satisfaction with your product. For example, a channel partner could provide the maintenance for your product if it were equipment. In the case of a hospital bed for a home care setting, the durable medical equipment (DME) store most likely delivers and sets up the bed. The selection of your channel partners will be one of the start-up's biggest decisions in terms of manifesting customer satisfaction and creating a brand to the customer as the channel partner becomes the "face" of your brand.

Looking at the *Manufacturer's* box in Figure 7.1, one will notice that there are four lines extending from that box. These lines represent the ways in which a manufacturer can get their product into the hands of their customer.

The first line demonstrates how the manufacturer may be able to sell directly to the retailers. An example could be in the category of medical surgical supplies, where surgical masks could be sold directly to a pharmacy chain such as Walgreens, CVS, or Rite Aid. Another example could be specific drugs (although most drug distribution goes through distributors).

The second line demonstrates that a manufacturer may decide that it is more appropriate to sell into a distributor because the cost of getting the product to the customer may be too high. For example, McKesson, Cardinal, Henry Schein, Owens & Minor, and others distribute to a majority of the health care providers. Looking at the *Alternate Site Provider* box in Figure 7.1, one can see the physician's office as a provider location. For an individual company that may sell only a handful of products into this space, it is difficult to afford a full-time sales professional to find the individual physician offices in which their product could be used. This would not be an issue for a distributor, who carries hundreds of thousands of individual SKUs. Many of us have gone into a physician's office building and noticed numerous physician types such as the gynecologist, the oncologist, the pediatrician, the endocrinologist, and others as tenants of the building. Let's assume that your company's product can only be sold to the pediatrician and the average pediatrician only needed five of your products in any given month. Is it affordable to pay a full-time sales professional to travel to this one physician's office building to only find that one pediatrician? The answer is probably not: this is the value of the distributor. As the distributor has a product for everyone in the building, it is affordable for them to pay for a full-time sales professional.

A third arrow from the manufacturing block is where the manufacturers sell directly to the hospitals. These tend to be higher-end products such as artificial hips, radiation treatment planning systems, stents, and other such products.

The fourth arrow from the manufacturing block is to the alternate providers. It should be noted that the alternate site provider block is not an all-inclusive block, nor should it be considered to be one category of customer. For example, a manufacturer could decide to sell directly to nursing homes and also decide to use a distributor for physician offices.

Before we leave Figure 7.1, it is important to note that products do not simply move from left to right. For example, pharmacy benefits management (PBM) companies distribute prescriptions and other health products for managed care plans, government agencies, and insurance companies. These companies buy in bulk. These companies can also provide services such as claims management. As it relates to Figure 7.1, PBM models can be derived from the *Private Insurer* box, the *Management Service Organizations* box, and the *Disease Management Organizations* box. This demonstrates that numerous permutations of channel partners can exist for the same product category. Top companies in the PBM space are Caremark, Aetna, Express Scripts, OptumRX, and Humana Pharmacy Solutions.[14]

Our last discussion on the numerous permutations of channel partners should be of the utmost concern to the start-up. Health care reform, the portability of electronic health care records, the creation of exchanges, social media tools and mobile applications, and accountable care organizations will all either reconfigure the value chain or move the power within the existing chain.

During the 2014 Innovations in Health Care Technology Conference held by Carnegie Mellon University (http://ihtconference.com/), the author hosted a panel discussion regarding the shifting of influence in the pharmaceutical industry. Historically, branded pharmaceutical manufacturers possessed the majority of the value chain influence or power. However, maturing markets are resulting in generics and the future of personalized medicine is resulting in more targeted drugs that break market sizes into smaller segments. The future business model does not appear to align with the past blockbuster drug business model where pharmaceutical companies generated significant sales channel influence. As power shifts, past servants can quickly become the masters. For example, organizations such as Cardinal Health Pharmaceutical can consolidate orders from hundreds of manufacturers into bulk and nonbulk deliveries for health care facilities and retailers. These products include generic, specialty, over-the-counter, and branded prescriptions. Their customers are broad, including hospitals, independent pharmacies, and retail chains. They offer management services and handle inventories, logistics, online procurement, and other administrative tasks. The value of their service to their customers may offset the historic brand influence of the blockbuster pharmaceutical company.[15]

The implication for the life science start-up company is additional risk, as health care will be undergoing continuous reorganization in terms of influence and power. New types of organizations may emerge and existing sales channel models may shift. A start-up with great technology may not be enough to ensure a liquidity event. The company must obtain the appropriate personnel skilled in health care sales channels and monitor the shifts in power. As a life science start-up can take several years to obtain market approval, a lack of guidance in this area could result in a product launching into an abyss.

Law Firms

A start-up company is most likely going to have one or more law firms. The start-up may have a law firm serving as general corporate counsel and yet another one focused on their intellectual property. The start-up's regulatory strategy might come from a specialized firm, while another may be focused on their reimbursement strategy. Although it might seem efficient to keep all your legal activities in one firm, it may not always be to your advantage.

The author had an opportunity to witness a startup company successfully plot a significant de-risking strategy by using multiple firms. The de-risking not only made sense from an operational perspective but it also allowed the start-up company to de-risk itself in the eyes of funders. The company's product was in a relatively new subcategory, however, the category itself had been around for 15 years. Although the subcategory was young, large corporations had staked a number of intellectual property claims within the category itself. This led investors to wonder if the company could successfully stake a claim to the subcategory that would keep those in the broader category out. Additionally, the broader category players had not been able to get into the subcategory because their products could not meet the FDA standards.

To alleviate anxiety among potential investors, the CEO hired the regulatory law firm that led the most recent FDA approval in his segment. That same law firm had tried to help another company navigate into that same subsegment. The company failed because its product could not meet the standard. In short, the CEO was able to demonstrate with the credentials of this law firm that his regulatory strategy was well advised.

The next concern was intellectual property and concern that the space was crowded. The CEO hired a different law firm that won the last intellectual property battle in this technology space. This was wise for numerous reasons: the firm was educated on the space by someone else's dollars and the CEO could demonstrate with the credentials of this law firm that his intellectual property strategy was sound.

Last, the CEO hired the law firm that got the last reimbursement codes from CMS. This law firm was able to validate the CEO's reimbursement strategy.

The CEO had a short-term and long-term objective associated with the strategy. As this company was a seed-stage company at the time, his short-term strategy was to demonstrate to funders that his strategies were sound based upon independent, third-party opinion leaders. His long-term strategy was associated with his exit strategy. These law firms were well known to, and trusted by, his acquirers. This would aid him during his final due diligence. Additionally, since these law firms had ongoing (nonconflicted) work with his acquirers, he was also counting on these firms sharing an occasional good word to his future acquirers. This is not to suggest the law firms would share anything confidential.

The author got to see the value of the strategy. The author worked in an incubator which frequently hosted venture capitalists who flew into the region to see multiple companies present. On this particular day, the company referenced above was to present to a venture capital organization. We will call this company 2 for clarity. Prior to company 2's presentation, another company (company 1) presented to the venture capitalist. Company 1 was actually more mature than company 2. The first company came in and presented their technology; the venture capitalist immediately started asking questions regarding the company's regulatory, intellectual property, and reimbursement strategies. Company 1 decided to go with one law firm and they hired a regulatory manager from an existing Fortune 500 company. As company 1 went about answering the VCs questions, the VC started to second-guess all the company's strategies, suggesting his experience and opinion was better than company 1's. The presentation was designed to last an hour and company 1 never got beyond the third PowerPoint slide: it was a complete disaster.

Enter company 2. As the venture capitalist did not have a pleasant experience with the first presentation, he did not let the CEO of company 2 get beyond his first slide without asking him his regulatory, intellectual property, and reimbursement strategies. Without even flipping slides, the CEO started to tell the story that was referenced above. There was no challenging or second-guessing the CEO because he not only had independent, arms-length validation of his strategy, he also had the players that fought the last battles in those segments and they were publicly known.

The CEO got through this part of his presentation in less than 20 minutes, got a smile from the venture capitalist, and jokingly asked if he (the CEO) could pass his first slide and tell the rest of the story.

What is the lesson from this story? The CEO recognized that he did not yet have the evidence at the seed stage to effectively de-risk his company. He took the time to recognize a strategy to overcome this obstacle. In selecting his firms, in the spirit of the book *The Tipping Point*, he hired Mavens and Connectors. In the short term, the Mavens provided expertise for funders. In the long-term, these Mavens are also connected to his future acquirers. His strategy laid the foundation of his future due diligence network.

Other Firms

In many segments of life sciences start-ups, there are service firms and marketing communications firms that have links back to acquirers. It is advisable to look at the suppliers, contractors, and channel partners of your future acquirers. Incorporating trusted members of the value stream may be of great value to the start-up. A caution is that these value stream members are more likely to be making more revenue from your acquirers than from your start-up. Confidentiality, prioritization, and other risks can also occur.

One channel partner that is frequently overlooked is in the marketing communications and/or public relations agencies. Using an acquirer's agency can have multiple benefits. The first is that the firm has a high pre-existing knowledge base, as the acquirers have invested in the agency's education: an education you do not have to pay for. The second benefit is that this agency may be able to help you reconcile and align your vision of your product. If your vision is inappropriately stated, the agency can easily help you align with the norms of this particular category. If the start-up does not pick an experienced agency and the same alignment is attempted, the start-up cannot be as confident in their recommendation as those of an experienced agency. If the issue is that the industry is not yet aligned to the new vision your product offers, the acquirer's agency is in the best position to help you develop a game plan to educate the industry. Finally, as the agency is most likely still connected to the acquirers, they are now part of the start-up's social network and can behave as a Connector and a Salesman in educating acquirers.

Customers

Building out your customer strategy should be more thoughtful than merely speaking to anyone that will listen. There are three aspects of your customer strategy that you should consider. The first is from the perspective of the start-up company testing its go-to-market model. The second is from the perspective of funders. The third is from the perspective of the acquirers.

The Start-Up Company Perspective

From the start-up company's perspective, a marketing and sales model needs to be created to forecast revenue. That model should be no less detailed than the process model used to manufacture a product. All the processes, supporting efforts, and anticipated time frames and yields should be predicted. As a start-up company launches, it needs to have a supporting measurement system to measure itself against its initial model so that problems can be anticipated and adjustments can be made at the earliest sign of variance.

Over time, the company will start to understand the time frame it takes to acquire a new customer, when to expect a reorder, and what sort of retention rate they should expect. Also, over time a start-up company will identify, with measurement, the differences in subpopulations within their markets. For example, in medical devices, frequently community-based physicians have historically adopted technologies faster than academic physicians. Yet medical device companies know that it is the academic physician's support that convinces late adopters and maximizes markets.

The Funder's Perspective

Many start-up companies limit their initial sales effort within a few hundred miles or so from their corporate headquarters. Although this is pragmatic from the start-up company's perspective, it does not give the funder a sense of the geographic acceptance of the product. Many funders are interested in a demonstration of the product's ability to be salable in the major regions of the country, for example, the East Coast, the Southeast, the Midwest, the West Coast, and the Southwest. Additionally, depending on the company's regulatory approvals, the funders may look for a demonstration of customer acceptance in Europe and/or Latin America.

Start-up companies are advised to understand the situation beforehand so that they can better balance sales force expenses with funder expectations.

The Acquirer's Perspective

As acquirers are frequently already selling products into the existing physician or provider area, they are often looking for their thought leaders and top customers to give a favorable opinion of the product.

In addition to the above, the acquirers may be interested in the opinions of physicians that support their competitor's products. The thought here is that if your products are favored, an acquisition could allow access to new customers.

Start-ups are advised to determine who their acquirer's thought leaders and/or top customers are and attempt to add them into their due diligence network.

DOMAIN-EXPERIENCED PERSONNEL

Hiring domain-specific personnel who are well versed in sales and marketing in a specific vertical is critical. Unfortunately, start-ups struggle with the transition from a development company into a commercialization company. Development personnel do not make as much money as domain-specific commercialization personnel. As a result, companies are tenuous about bringing these people on too early because of the expense. All too often, the start-up company attempts to hire the personnel as close to the launch date as possible. Although companies understand that they need to hire their R&D, quality, and operations people long before they would expect to produce one unit they somehow do not understand that sales and marketing personnel need time to plot their processes well in advance of launch. This hesitation frequently leads to the first-year revenue forecast being missed. This can become a problem, as acquirers look to first-year projections to determine customer acceptance. Missing those early projections can have a great impact on early acquisition, and, in the long run, this is more costly for the company than if they had brought the talent on earlier.

ACQUIRERS

We will expand upon the topic of acquirers in the management team section. However, one is advised to not only understand who the acquirers are but to also understand the strategic plans of each acquirer. Additionally, one should understand each acquirer's business model and anticipate how the start-up's business model would fit within the acquirer. Anticipation of business model fit must go beyond generalities. Although the author appreciates that start-ups have constraints in resources and time, in a perfect world the start-up would not only understand the gross margins in general profitability of each acquirer, they would also understand the critical processes.

An example could be illustrated here. There was a startup company that had designed a material that brought more natural colors into contact lenses for the market where people desire to use contacts to change the appearance of their eye color. The product worked beautifully and presented a more realistic eye color than existing products. Unfortunately for the start-up company, they did not recognize until they were well underway with the project that they would never be acquired by existing contact lens manufacturers. One of the major contact lens manufacturer's marketing people saw the product and were delighted. Marketing people anticipated that this could be a new major market segment for the company. When the marketing people brought the start-up company in to meet with the manufacturing department, the marketing people and the start-up realized that the vision the start-up company had was not possible. The issue was that the material used to make these contact lenses was not something that any current lens manufacturer could use in their existing manufacturing processes. It was quickly determined that if the acquirer created a process that could utilize the new lens material, the cost of the lens would be well beyond what existing market prices would bear.

This example demonstrates the lack of seeking domain-specific knowledge. With forethought, this information could have been readily understood before the start-up company even began their design process. As a result of the lack of domain-experienced personnel, a significant financial loss occurred.

CAPITAL PROVIDERS

In Chapter 6, we discussed in great detail the rules of the individual capital providers: friends and family, angels, venture capital, corporate venture capital, bankers, and the value of an IPO. We are not going to go much further with this topic here, except to say that these capital providers each have their own reputations with other capital providers and acquirers.

As one considers understanding the start-up's acquirers, a review of their M&A history should be performed. That analysis should go far enough back to understand the capital providers that were involved in the companies that were acquired. This list of capital providers and individuals would be a great starting point in plotting your preferred capital provider network.

The benefit of this would be that the acquirers would trust these funders and the funders would most likely be what is referred to as "smart money." Smart money is when the capital provider not only provides money but also its relationships as well as its expertise. Unless you have previous relationships with capital providers, it is hard to determine whether they are smart money, a track record of leading exits toward acquisition is certainly an indicator.

For seed and early-stage companies, they would be looking for funders that have relationships with these individuals. As stated earlier, people like to invest with people they trust and have worked with before. Understanding this network is not as difficult as you may think. Databases such as Venture Source are available to create this analysis. Think of the concept of LinkedIn. Have you ever searched for an individual on LinkedIn and then realized that they had a relationship with someone else in your network? LinkedIn helps you understand that network and provides you an infrastructure so that your network can make an introduction to the person whom you are targeting. You are really trying to do the same thing in your capital provider analysis.

SUPPLIERS AND CONTRACTORS

Returning to Figure 8.1 in Chapter 8, we recognize that one of the major inputs into a start-up is its network of suppliers and contractors. When defining a domain-experienced network, suppliers and contractors can be invaluable to the start-up company.

On the positive side, a well-selected supplier or contractor can allow the start-up company to leverage the existing infrastructure of that firm. For example, many medical device start-ups will not have the funding to create an appropriate design control system that supports the construction of the medical device's initial specifications. This is required by the FDA. By utilizing an outside design firm that already has a significant investment into an appropriate design control system, the start-up firm can use that infrastructure to support its regulatory documentation. It would be difficult for a seed stage start-up to both find and afford the talent to create such a support system—not to mention the time to create the documentation system.

Another example is a start-up pharmaceutical company. All pharmaceutical companies must produce their initial product within a manufacturing environment that meets all of the pharmaceutical regulations. This is a detailed and expensive infrastructure and its cost would be prohibitive for a start-up. Not only is utilization of an outside manufacturer more cost-effective, it also ensures investors that an appropriate infrastructure supports their investment. It de-risks the program to the outside investor. As importantly, this strategy may also be appropriate to ease acquirer's concerns. Some acquirers are comforted by the thought that a skilled supplier is responsible for the production of the product they are acquiring, as the acquirer knows that the contractor desires to continue the business post-acquisition.

Today in health care information technology businesses, cloud and outsourced IT infrastructure not only ensures a cost-effective scale but also ensures that the latest in rapidly evolving security techniques are deployed. Importantly, outsourcing IT infrastructure comforts the acquirer that an easy transition from start-up to acquisition is almost ensured.

Another benefit for funders and acquirers is that suppliers require agreed-upon specifications in order to be able to deliver on expectations. This need demands a well-thought out process and appropriately detailed level of documentation to be successful. This documentation has the benefit of enforcing discipline; however, it can also have its risks.

The necessity of rewriting specifications for each revision can arguably get in the way of rapid prototype or "beta" cycles. A beta cycle is when the product is developed, presented to the customer for feedback, and either accepted or reworked. Should rework be in order, another cycle must be begun. Why spend the time documenting something that is not going to ultimately be accepted?

Rework cycles can be costly and time-consuming. For example, in one medical equipment company the CEO invested in a small tool shop. The tool shop was used for rapid prototyping. Unfortunately for the CEO, he planned his fund-raising activities in a way that left him financially short to achieve his next fundable milestone. The board of directors decided that they did not want to risk a down round and decided that cost-cutting was the appropriate approach to bridge the company to the next fundable milestone. Unfortunately for the CEO, he could not convince his board to keep his tool shop employee and a layoff resulted. Within 60 days, product development delays were massing to the point where the fundable milestone was moving further away, not closer. The reason for the delays was that when the design engineers went out to the customer and had design changes, those design changes required documentation. The documentation was then sent to the supplier, who would either accept the changes or propose another method meeting specifications. These iterations were time-consuming and arguably wasteful if the customer did not accept the proposed change. In the time that it would take to do these iterations with the supplier, the company's tool shop could have made several changes to the physical product and brought it to the customer for feedback. Within 60 days, the board of directors re-authorized the CEO to hire that resource back.

Selecting the wrong supplier can also add great risk to the start-up if proper due diligence is not performed. There is the obvious loss of time and money associated with making a poor selection. However, there are other costs that can be more debilitating to a start-up company.

In one example, a start-up company did not have the appropriate procedures in place with their supplier. During the prototyping and clinical trial phase, the relationship went flawlessly. However, as the company received FDA approval and started their launch, they had numerous customer complaints. Upon investigation, the start-up company discovered that their supplier outsourced its scale production to China. Not only were there production issues caused by this change, the company's intellectual property had been now sent to a country that was notorious for not respecting intellectual property rights. In the marketplace, instead of delighting their customers and demonstrating customer acceptance to acquirers, the company was demonstrating concern to both their customers and to onlooking acquirers.

In another example, a start-up's supplier was involved with another start-up company—we will call this start-up number 2. What was unknown to the start-up company was that start-up 2 was involved with and eventually acquired by the start-up's future competitor. Once the competitor purchased start-up 2, it demanded the supplier to cease work with the start-up, resulting in a significant loss of time and money. This story speaks to the importance of building your social and due diligence networks. On the one hand, the story above represents the risks associated with the supplier. That same story also speaks to opportunity.

If the start-up had thoroughly considered that their supplier was a part of their network, they would have recognized the potential for the problem and could have created legal paperwork to protect themselves. The same example expresses that suppliers in a particular vertical most likely have relationships with competitors and potential acquirers. Moving into these relationships with this understanding would allow the start-up company to plot strategies to minimize risk and maximize the network value of the supplier relationships.

Our last discussion is on the pros and cons of contractors. Contractors can allow an organization to bring in domain-specific talent for the short period in which they are needed. In many cases, the contractors have worked for the start-up's future acquirers. This provides a great opportunity to ensure that the start-up is aligning its business with the needs of acquirers. Additionally, contractors provide independent validation of the board of directors and funders that internal activities are going according to plan.

The downside of contractors is that their connections, networks, and future contracting assignments come from the acquirer's network. The start-up company must have appropriate confidentiality agreements in place with its contractors to ensure that its technology and strategies do not leak out of the organization.

THE MANAGEMENT TEAM

Ultimately, the due diligence and social network in Figure 8.1 in Chapter 8 is designed and created by the management team. Before we begin our discussion, let's define some terms.

In general, when people refer to the management team they are talking about the CEO and his direct reports. However, for our discussion we are going to call this the *management board*. The reason for this is that, in a start-up company, a board of directors and the board of advisors are also critical in establishing the network strategies.

THE MANAGEMENT BOARD

The management board consists of the CEO and his direct reports. This team is critical to the success of the organization and single-handedly accountable for building the due-diligence and social network.

It is important for the CEO to consider his due diligence and social network prior to hiring any direct report. We have spoken of the importance of domain experience in creating the management board. Let's take a moment to expand the meaning of domain experience. Thus far we have used domain experience to suggest that the individuals should be from the start-up company's vertical/domain. However, domain experience should also be considered as having relevant experience to the task at hand. For example, for a drug/device combination, who do you pick? For a medical device/IT product, do you pick someone with device expertise or IT expertise?

The ultimate answer lies in balancing out the team. For example, the author would be inclined in a drug/device combination to hire a regulatory leader from the drug industry and to support that individual with someone in the medical device industry. There is no one right answer except that in start-ups that are combinations, the CEO must be aware that hiring from only one dimension will ultimately be devastating. At the same time the CEO must also be aware of the need to manage to completely different cultures. The author has been involved in start-ups that traverse customer verticals such as dialysis, vascular surgery, and interventional radiology as well as being involved in companies that traverse medical devices and drug technologies.

In the case of traversing customer verticals, philosophical conflicts play out in the theater of the customer. Unless challenges are anticipated and the appropriate people hired beforehand, a lack of clarity occurs in the positioning of the company. Recognizing and actively managing the different cultures allows for a synergistic approach where 1 + 1 = 3. Not managing the culture, where differences are discussed and reconciled, results in conflict. In one start-up in which the author was involved, this conflict was so great that it resulted in the bankruptcy of the company.

The value of building a team that successfully traverses specialties can result in a new business model. The team and the business model can become as valued an asset as the product itself. If done correctly, acquirers will look at the business model and team in valuing the acquisition price of the company.

The Board of Directors

The board of directors is as critical to the company as its management board. The start-up must be very mindful of whom they add to the board of directors, particularly when selecting the chairman of the board. Once installed, board members are frequently very difficult to move off the board. The chairman of the board is particularly a critical selection as he presides over the board meetings and can act as the company's representative to the outside world.

In the spirit of our social network discussion, it is important for the start-up company to understand why each member of the board is selected and what their motivations are for participation. In some cases, an early investor demands a seat on the board and the willingness to take that investor's money must take into consideration the impact of having him on the board. Practically speaking, seed stage companies seldom can do much about this situation, given the difficulty in finding pre-seed and seed financing. In situations such as this, it becomes critical for the founders and the CEO to think about their long-term board strategy.

For example, if this initial investor does not have the appropriate domain experience, the start-up risks the board member providing ill-advised direction. Although many times these early investors do not possess the funds to continue to hold their seats in the long run, thoughtful strategies can be deployed. These strategies start with the company's governance, which generally is first put into place by the founders of the company. For example, let's assume you are in a situation such as the one described above: what can be done?

One solution could be that you would start out the board of directors with three people. One position could be held by the CEO, and the CEO could also be the chairman. The second position could represent the founders, with the third position being held by the investor. If the CEO is also a founder, and he decides who to place in the founder's board seat, the CEO continues to have great influence in the company.

In another situation, the board was expanded to five to accommodate two desired Series A investors. It was known that one of the Series A investors would not continue to invest in future rounds. In this case, both investors had domain experience and were valued. The company's strategy was to start out with the board consisting of the CEO, two positions representing the founders, and two positions representing series A investors. When the Series B round would occur, given the size of the round, two positions were required for Series B investors. As the company was thoughtful in their board strategy, the high-net-worth Series A investor also invested in Series B and he continued on the board. However, he no longer represented Series A but moved his board seat to represent Series B. The other Series A member continued on the board. The founders' board seats were reduced from two representatives to one. The founder's seat was now added to represent the Series B investor.

As is now known by the reader, the author is associated with an incubator. The situation above was the start-up company trying to protect itself from control by others without domain experience. Frequently the opposite occurs where the founders are not capable of managing the company long term. These founders can try to create board strategies designed solely to allow them to maintain control. This founder approach is not about maximizing the value of the company; it is about maximizing their personal control. In this situation, as an incubator, we try through our investment to influence the board structure. We would first ask that the CEO position be separated from the chairman position. Next, we would ask for a board seat that represents our pre-seed or seed investment. This sets up a situation in which the board chairman must be agreed upon by both parties. This allows the company to start off on a more objective footing. When it comes to the next round, as an incubator we are in a position to either continue on the board, expanding the board to five members, or we can step off the board and allow a new investor to take our seat.

Returning to Figure 8.1 in Chapter 8 for a moment, we recognize that all of our previous discussions were based upon poor representation solely from the capital provider portion of a network diagram. This speaks to the scarcity and importance of funding to the launch of a successful start-up. However, other members could be extremely valuable to the board. For example, a representative from the acquirer's industry could ensure domain alignment. They would offer guidance as to what needs to be done to trigger an acquisition. Obviously, this is critical guidance for the start-up company.

Last, an executive who represents the customer's industry, such as an insurance provider or a health care provider, can be a valued resource on a board of directors. There is frequently confusion as to what customer representation should exist on the board of directors and what representation should be moved to the board of advisors. It is the author's opinion that this shifts with the life of the company. Having an executive level board member is valuable from the customer's perspective. One could appraise the value of a president or vice president from a hospital or insurance plan participating on your board.

Typically, customer representation comes from an individual physician and that may be appropriate in the short term, as they ensure that the board does not deviate from understanding the customer's problem. However, as the company advances beyond prototype, board conversations shift to funding and exit strategies. These clinicians have little to offer in this environment. Additionally, based upon the extent of the physician's organizational conflict-of-interest policies, selecting a physician for your board may be a decision that would inhibit the physician's institution from buying your product.

The Board of Advisors

Should you have a physician on your board during your preseed and seed stages, a respectful transition is to move them onto your board of advisors. In some cases, additional respect can be shown by appointing the physician to the chairmanship of the board of advisors. As boards of advisors frequently have no material financial interests in a company, conflict of interest is less of a concern.

The first question for the start-up is whether you want a formal or an informal board of advisors. The value of a formal board of advisors is that it is a statement to funders, customers, and acquirers. The next question, should you choose a formal board of advisors, is whether your rationale for creating them is for market research reasons or as a sales tool. A formal board of advisors allows the management board and/or the board of directors to have a more formal give-and-take discussion with an organized body. By having a formal process, the give-and-take discussions are structured more like formal market research.

Some start-ups view their board of advisors as assisting the sales process. To this type of start-up, the board of advisors represents advocates and provides personal testimony to accelerate sales adoption. There is nothing wrong with this strategy as long as the company is doing it purposefully.

In the end, many companies use a mixed strategy; however, it is the formalized approach that is important as it better contextualizes the feedback and insight than an informal process.

Many start-ups that choose an informal approach are only interested in the advisors advancing the sales process. Like utilizing LinkedIn, they are simply creating a public network for sales expansion. Again, there is nothing wrong with this, as long as the start-up is aware of the fact that the informality of the process risks receiving advice out of context.

A THOUGHTFUL GATHERING OF EXPERIENCE REDUCES RISK

As a child, did you ever get advice on how to do something or handle a situation and ask yourself, how did they know that? Perhaps you have had an instructor who navigates the complex with ease causing you to wonder, how long did it take him to learn that?

According to K. Anders Ericsson, professor of psychology at Florida State University, it takes about 10,000 hours of practice to become a subject expert. From a professional perspective, if you assume that there are 40 hours in a workweek, and 50 workweeks a year, there are 2,000 professional hours a year available for mastery. That would mean mastery could average five professional work years or more (10,000/2,000). Zach Hambrick, associate professor of psychology at Michigan State University, designed studies to validate the 10,000 rule, and found other factors such as genetics also contributed to mastery.[16]

The point of this entire chapter was to demonstrate the value of gathering domain-experienced personnel to reduce the risk profile of the company. Utilizing Figure 8.1 in Chapter 8, a network module was presented to demonstrate the complexity of the process. Given the number of network modules and the number of functional positions within each module, it is unlikely that any one human being can achieve mastery: both the number of hours available in a career and the genetic makeup of any particular individual limit this.

What can be achieved, though, is a thoughtful process for gathering the appropriate domain experience, as demonstrated by Figure 8.1. It is important for the management team to agree upon such a process and work collaboratively to acquire the appropriate network components. Adding more complexity to the challenge is that individuals' traits must be balanced as one collects and builds the network. This balance is both at the personality level (for example Connectors, Mavens, and Salesmen), and also at the constituency level, with funders, customers, and acquirers aligned.

We end this discussion with a new understanding that the due diligence business process management system (see Figure 9.1 in Chapter 9) is a subledger to the company's social network creation. The company's social network (see Figure 8.1 in Chapter 8) is a subledger supporting the company's brand plan. All this planning is initiated toward that magic moment when you create an environment where an idea crosses a threshold, causing it to tip and spread like wildfire.

DISCUSSION QUESTIONS:

1. Explain the concept of "six degrees of separation."
2. What are some benefits of "connectors"?
3. What do "thought leaders" bring to the table in terms of start-up companies?
4. How did Paul Revere's story relate to being a Connector and a Salesman?
5. What is "burn rate"?
6. What are "mavens"?
7. Is the Salesman essential in realizing the expertise of the Maven? Why or why not?

8. Differentiate a patent from a trademark.
9. What are Channel Partners?

NOTES

1 *Wikipedia*, s.v. “Frigyes Karinthy,” http://en.wikipedia.org/wiki/Frigyes_Karinthy (accessed March 15, 2014).
2 Malcolm Gladwell, *The Tipping Point* (New York: Little, Brown and Company, 2006).
3 Practice Management, Types of Business Entities, *American Speech Language Hearing Association*, http://www.asha.org/practice/BusinessEntities/ (accessed March 15, 2014).
4 Stephanie Paul, LLC or LP: What’s Best for Your Business? *LegalZoom*, February 2011, http://www.legalzoom.com/business-management/starting-your-business/llc-or-lp-whats-best.
5 Ibid.
6 Choosing Your Business Structure: S Corporation, *U.S. Small Business Administration*, http://www.sba.gov/content/s-corporation (accessed March 15, 2014).
7 Rick Nauert, Why First Impressions Are Difficult to Change: Study, *Livescience*, January 19, 2011, http://www.livescience.com/10429-impressions-difficult-change-study.html.
8 The Coca-Cola Company, Coke Lore: Trademark Chronology, *Coca-Cola Journey*, January 1, 2012, http://www.coca-cola-company.com/stories/coke-lore-trademark-chronology.
9 Mallory Russell, How Pepsi Went from Coke’s Greatest Rival to an Also-Ran in the Cola Wars, *Business Insider*, May 12, 2012, http://www.businessinsider.com/how-pepsi-lost-cola-war-against-coke-2012-5?op=1.
10 Update: Exploring FDA approval Pathways for medical devices, Mass Device, Danielle Kirsh, September 12, 2019, https://www.massdevice.com/exploring-fda-approval-pathways-for-medical-devices/#:~:text=The%20average%20cost%20to%20bring,average%20costs%20of%20%2494%20million. (Accessed February 2021).
11 Our Services, *BSI Group*, http://medicaldevices.bsigroup.com/en-GB/our-services/ce-marking/.
12 Certification, ISO Management System Standards, *ISO*, http://www.iso.org/iso/home/standards/certification.htm.
13 Drug Applications and Current Good Manufacturing Practice Regulations, *U.S. Department of Health and Human Services, U.S. Food and Drug Administration*, http://www.fda.gov/Drugs/DevelopmentApprovalProcess/Manufacturing/ucm090016.htm.
14 Updated: The PBMs by Market Share, Becker Hospital Review, Alia Paavola, May 30, 2019 https://www.beckershospitalreview.com/pharmacy/top-pbms-by-market-share.html, (accessed February 2021).
15 Cardinal Health Pharmaceutical, *Hoovers Online, A D&B Company*, 2014, http://subscriber.hoovers.com.proxy.library.cmu.edu/H/company360/overview.html?companyId=104517000000000.
16 Maria Szalavitz, 10,000 Hours May Not Make a Master after All, *Time Magazine*, May 20, 2013, http://healthland.time.com/2013/05/20/10000-hours-may-not-make-a-master-after-all/.

14 Determine the Acquirer's Strategic Future and Purchase Triggers

FIND YOUR TARGETS (POTENTIAL ACQUIRERS)

Let's start with a brief calibration tangent. In marketing, targeting a customer first requires an analysis of all potential buyers for your product. Next, all the buyers are segmented by their particular desires or interests to form your initial target market. It is important not to market to the entire population, because if your product does not meet the needs of these buyers, it would be financially wasteful to pursue them. Those segmented buyers whose interests are matched by your product become your initial target.

Before descending on your target, a supporting analysis is performed; that is called the *cost of customer acquisition*. The cost of customer acquisition is calculated by accumulating the total cost to gain customers. Most people think of this as simply the cost of sales representatives, however, it is inclusive of the cost of clinical support and other field activities necessary to gain and retain the target. Return is then calculated by taking the revenue gained from the effort and subtracting it from the total cost to determine if the effort generates sufficient returns. If the effort does not generate sufficient returns, the process should be reworked or the effort should be abandoned.

If the process does generate sufficient returns, frequently the targets are again segmented by return and time-to-close. If this can be determined, it is natural to select those that provide a higher return and buy more quickly than those who buy more slowly. All else being equal, one can intuitively expect these customers to have a less expensive cost of customer acquisition. So, why are we having a conversation regarding cost of customer acquisition, when our topic is the acquirer's strategic future and purchase triggers?

The reason for this discussion is that the start-up CEO must understand their potential acquirers' strategic future and purchase triggers. All the potential acquirers should be segmented by strategic interest and by purchase triggers. Understanding this information allows us to separate the entire acquirer population into segments. The start-up's first sort would be by those strategically interested in our investment property (start-up). Next, we would follow this with estimating the time it would take the acquirer to acquire. In our discussion, we will call this an *acquirer's purchase trigger*. Our next natural sort would be by purchase trigger.

DOI: 10.1201/9780367533052-14

For example, if our start-up is of equal interest strategically to an acquirer, one acquirer may be more aggressive and buy upon the issuance of a patent, while the second acquirer may buy only upon achieving $5 million in revenue. Clearly, until sufficient revenue is generated, spending material effort in obtaining an acquisition from this company before revenue is wasteful. So how do we go about this process?

Develop Industry Micro and Macro Maps

The first step to segmenting requires us to understand where a potential acquirer is placed in the market. As most entrepreneurs generally know the products that are sold into their targeted market, developing a microsegment overview (map) is the easiest place to start. Looking at a micro map, Figure 14.1, the preparer should start by listing the competitors in their target market by row. For each competitor, create a column representing the product groups in which they participate. Defining the product groups may be the most challenging part if you do not have domain experience, however, the important point is to categorize them in a way that makes sense to you. In general, the major segments are collected by either physician type, call point (department or group), or some other method of your choosing. Using physician specialists in medical devices is intuitive because one can start to list the tools (products) that the physicians utilize in their daily practice. In our example, the micro map, Figure 14.1 has been populated with the major tools used in the cardiac rhythm management marketplace. The start-up is going to need to use their judgment in defining the micro map. For example, in a pharmaceutical setting, the micro map may be based upon a particular disease such as cancer, with the various treatments representing the columns. To see examples of different types of *micro map* examples by market vertical such as pharmaceutical, diagnostics, biotech tools, and health care IT, go to the book's Web site at: https://healthcaredata.center/healthcare-delivery/the-supplier-view/. The author updates these maps every 12–24 months.

Once the micro map gives the user a perspective of his individual segment, we then turn to the macro map, Figure 14.2, to determine how the company fits into a broader context, in this case a hospital. The macro map starts with creating a column from the micro map segment and creating a row for each competitor from the micro map. Now, continue to build out according to the appropriate columns. In this case, the author has used the columns appropriate to the medical device industry in the hospital. To see samples of different types of *macro map* examples by market vertical such as pharmaceutical, diagnostics, biotech tools, and health care IT, also go to: https://healthcaredata.center/commercializationstartup.

Now that the macro map is complete, the next question is, do any of the macro map companies serve another segment of the health care system? Starting with the Health Care Systems Flow Chart in Figure 7.1 in Chapter 7, determine where each potential acquirer participates in the greater market. Depending on the company you are looking at, their participation could be very broad. For example, Johnson & Johnson sells medical devices, pharmaceuticals, and diagnostic and consumer products. McKesson, perceived as a pharmaceutical distributor, also distributes medical-surgical supplies, has numerous health care service organizations, and has a significant health care information technology business. Once you understand

Cardiac Rhythm Management Micromap

Companies	Implantable Cardiac Defibrillators		Atrial Fibrillation (Afib)	Pacemakers	Replacement Heart (LVAD)	Catheter Cardiac Ablation	Intra Aortic Balloon Pump	Automated Implantable Cardioverter - Defibrillator (AICD) (tachyarrhythmia)	Electrophysiology	Cardiac Resynchronization Devices	Cardiac Monitoring	Cardiac Assist Devices	External Defibrillators
	Fibrillator Electrodes	Defibrillators											
Abbott		✓		✓	✓	✓				✓	✓		
Abiomed					✓						✓	✓	
Acutus Medical			✓			•							
Angiodynamics						✓							
Arterial Remodeling Technologies												✓	
Asahi Intecc Co.						✓	✓						
Asahi Kasei Inc											✓	✓	✓
Biosig Technologies									•				
Biotronik				✓				✓	✓	✓	✓		
Boston Scientific		✓		✓	•	✓		✓		✓			
Cardiac Science											✓		✓
Cardinal Health													•
Cath Rx			✓			✓							
Edwards Life Sciences					✓								
GE healthcare									✓				
Getinge Group							✓						
J&J						✓			✓				
Medtronic	✓	✓	✓	✓	✓	✓ ✓ •		✓	✓	✓	✓	✓	✓
Microport Scientific Corp.		✓	✓	✓						✓	✓		
Philips Healthcare											✓		•
Sichuan Jinjiang Electronic Science and Technology (JJET)						✓			✓				
Siemens Healthineers									✓				
Syncardia Systems					✓								
Teleflex							✓						
Welch Allyn											✓		

Legend:	
✓ →	Marketed Product
• →	New Addition

FIGURE 14.1 Micro map. (From Jordan, 1996.)
Source: https://healthcaredata.center/healthcaredelivery/the-upplier-view/medical-devices/

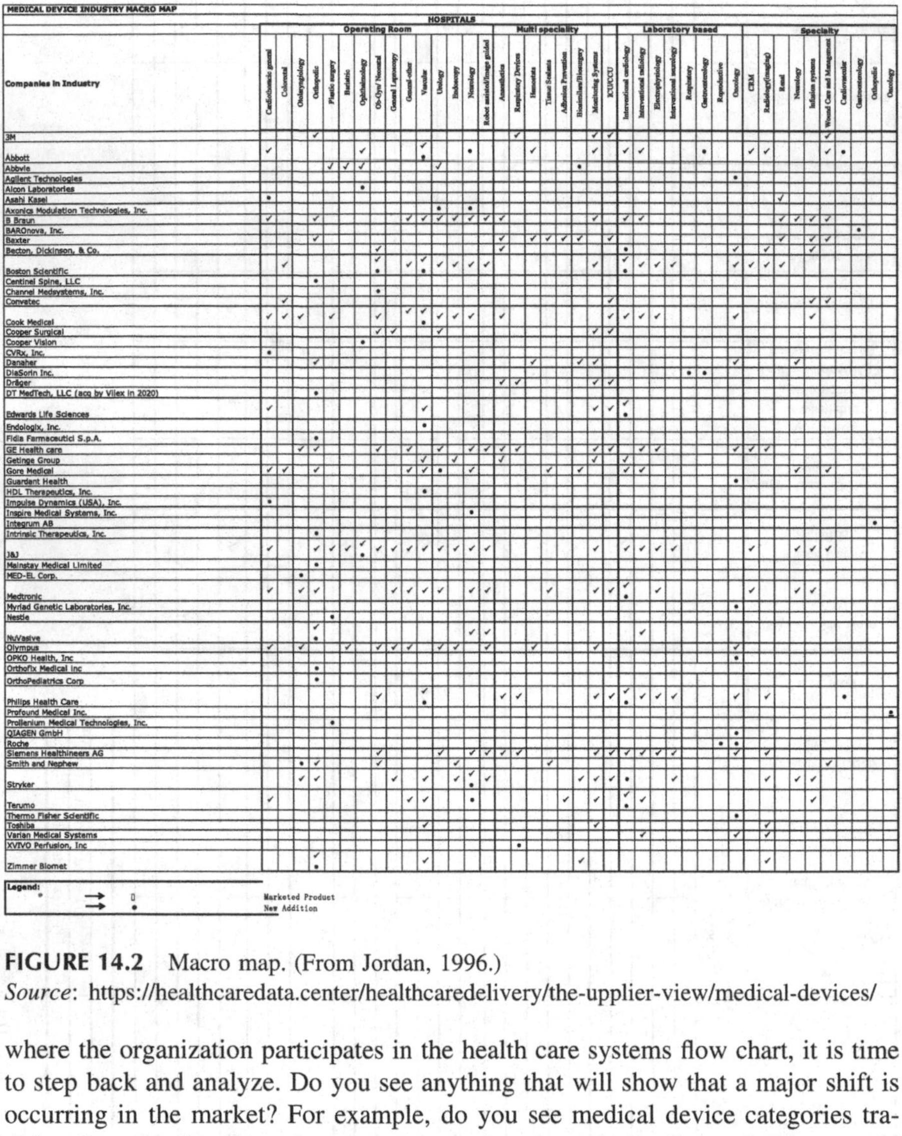

MEDICAL DEVICE INDUSTRY MACRO MAP	HOSPITALS																																								
	Operating Room															Multi speciality								Laboratory based								Specialty									
Companies in Industry	Cardiothoracic general	Colorectal	Otolaryngiology	Orthopedic	Plastic surgery	Bariatic	Ophthalmology	Ob-Gyn/ Neonatal	General Laproscopy	General-other	Vascular	Urology	Endoscopy	Neurology	Robot assisted/Image guided	Anesthetics	Respiratory Devices	Hemostats	Tissue Sealants	Adhesion Prevention	Biosimilars/Biosurgery	Monitoring Systems	ICU/CCU	Interventional cardiology	Interventional radiology	Electrophysiology	Interventional neurology	Respiratory	Gastroenterology	Reproductive	Oncology	CRM	Radiology(imaging)	Renal	Neurology	Infusion systems	Wound Care and Management	Cardiovascular	Gastroenterology	Orthopedic	Oncology
3M				✓													✓					✓	✓														✓				
Abbott	✓						✓				✓•			•				✓				✓		✓	✓				•			✓	✓				✓	•			
Abbvie					✓	✓	✓					✓									•																				
Agilent Technologies																															•										
Alcon Laboratories							•																																		
Asahi Kasei	•																																	✓							
Axonics Modulation Technologies, Inc.												•		•																											
B Braun	✓			✓						✓	✓	✓	✓	✓	✓	✓						✓		✓	✓									✓	✓	✓	✓				
BAROnova, Inc.																																							•		
Baxter				✓												✓		✓	✓	✓	✓		✓											✓		✓	✓				
Becton, Dickinson, & Co.								✓								✓								•							✓		✓			✓					
Boston Scientific		✓						✓•		✓	✓•	✓	✓	✓	✓							✓		✓•	✓	✓	✓				✓	✓	✓	✓							
Centinel Spine, LLC				•																																					
Channel Medsystems, Inc.								•																																	
Convatec		✓																					✓													✓	✓				
Cook Medical	✓	✓	✓		✓			✓		✓	✓•	✓	✓	✓		✓							✓	✓	✓	✓					✓					✓					
Cooper Surgical								✓	✓			✓										✓	✓																		
Cooper Vision							•																																		
CVRx, Inc.	•																																								
Danaher				✓														✓			✓	✓									✓				✓						
DiaSorin Inc.																												•	•												
Dräger																✓	✓					✓	✓																		
DT MedTech, LLC (acq by Vilex in 2020)				•																																					
Edwards Life Sciences	✓										✓											✓	✓	✓•																	
Endologix, Inc.											•																														
Fidia Farmaceutici S.p.A.				•																																					
GE Health care			✓	✓				✓		✓		✓		✓	✓	✓	✓					✓	✓		✓	✓					✓	✓	✓								
Getinge Group											✓		✓			✓						✓		✓																	
Gore Medical	✓	✓		✓						✓	✓	•		✓					✓		✓			✓	✓										✓		✓				
Guardant Health																															•										
HDL Therapeutics, Inc.											•																														
Impulse Dynamics (USA), Inc.	•																																								
Inspire Medical Systems, Inc.														•																											
Integrum AB																																								•	
Intrinsic Therapeutics, Inc.				•																																					
J&J	✓			✓	✓	✓	✓•	✓	✓	✓	✓	✓	✓	✓	✓							✓		✓	✓	✓	✓					✓			✓	✓	✓				
Mainstay Medical Limited				•																																					
MED-EL Corp.			•																																						
Medtronic	✓		✓	✓					✓	✓	✓	✓		✓	✓				✓			✓	✓	✓•		✓						✓			✓	✓					
Myriad Genetic Laboratories, Inc.																															•										
Nestle					•																																				
NuVasive				✓•										✓	✓										✓																
Olympus	✓		✓			✓		✓	✓	✓		✓	✓		✓			✓				✓									✓										
OPKO Health, Inc																															•										
Orthofix Medical Inc				•																																					
OrthoPediatrics Corp				•																																					
Philips Health Care								✓			✓•					✓	✓					✓	✓	✓•	✓	✓	✓				✓		✓					•			
Profound Medical Inc.																																									•
Prolienium Medical Technologies, Inc.					•																																				
QIAGEN GmbH																															•										
Roche																														•	•										
Siemens Healthineers AG								✓				✓		✓	✓	✓	✓					✓	✓	✓	✓	✓	✓				✓		✓								
Smith and Nephew			•	✓				✓					✓						✓																		✓				
Stryker			✓	✓					✓		✓		✓	✓•	✓						✓	✓	✓	•			✓						✓		✓	✓					
Terumo	✓									✓	✓			•						✓		✓		✓•	✓											✓					
Thermo Fisher Scientific																															•										
Toshiba											✓											✓											✓								
Varian Medical Systems																									✓						✓		✓								
XVIVO Perfusion, Inc																	•																								
Zimmer Biomet				✓•							✓										✓												✓								

Legend:
→ ▯ Marketed Product
→ • New Addition

FIGURE 14.2 Macro map. (From Jordan, 1996.)
Source: https://healthcaredata.center/healthcaredelivery/the-upplier-view/medical-devices/

where the organization participates in the health care systems flow chart, it is time to step back and analyze. Do you see anything that will show that a major shift is occurring in the market? For example, do you see medical device categories traditionally sold directly to hospitals now going through distributors? What would that mean to you? Should there be a scenario where you deem it necessary to look into another part of the health care system, for example, surgery centers; the micro and macro maps should also be created to look for other acquirers.

The value of this exercise is that one may uncover potential acquisition targets that might not have been traditionally thought of. For example, a micro map exercise of the orthopedics industry would not have predicted that Johnson & Johnson bought DePuy for $3.7 billion in 1998.[1] However, if one created a macro map prior to 1998, they would have noticed that orthopedic companies were purely isolated and did not participate in other macro map areas. In 2017, Bedcton Dickinson (BD) acquired C.R. Bard for $24 billion, why? An article in MedCity News stated that it was important for BD to move beyond its focus on diabetes to expand its disease states. Importantly, by having a broader product offering BD could be more competitive in bidding to hospitals who are seeking to consolidate vendors and improve pricing.[2] Looking at a macro map, one would have recognized if they wanted leadership in the operating room, a future orthopedic company would most likely need to be acquired. This exercise allows for those considerations.

ANALYZE THE INDUSTRY'S PRODUCT LIFE CYCLE

Figure 5.2 in Chapter 5 represents the product life cycle curve. The product life cycle curve helps one envision how markets evolve over time. Acquirers are motivated to seek innovation; to a marketing specialist they are looking for product that either:

- Creates a new, differentiated, and protectable market category, or
- Collapses the value steps or improves efficiency in an existing category, resulting in decreased cost or an increased benefit.

The product life cycle curve demonstrates that as demand declines, industry consolidation is necessary to adjust supply with demand while maintaining declining pricing.

At the *introduction stage* of the product life cycle, the product is initiated into its market. The priority at this time is to create awareness of the offering through promotional efforts. It is noteworthy that frequently competitors are few or non-existent at this point. If there are competitors, promotional efforts are oriented toward growing the category to the benefit of all versus competitive attacks.

At the *growth stage*, sales are generally escalating due to product awareness. This growth, with its corresponding cash, attracts competitors. Incumbent companies reinvest their cash windfalls to maintain position and hold off competitors. It is at this stage that promotional efforts change focus toward differentiating from the competition.

At the *maturity stage*, the market is established and sales tend to plateau: weaker competitors struggle for market share and ultimately leave the market and the remaining players intensely compete for market share. At this point, large cash profits are available for investors as product reinvestment is not as attractive based upon the market maturity.

At the *decline stage*, sales growth ceases and there are fewer players in the marketplace. As demand falls, companies either eliminate product lines or seek to extend life spans through new product line extensions or repositioning the products to new markets.

Perhaps some examples would be useful in sharing the importance of understanding the product life cycle. The first example is in the market introduction phase—the electric car. It is generally understood that this market lacks supporting infrastructure. Lacking market infrastructure constrains growth and the manufacturer's ability to sell features that would be desirable in later phases. For example, a manufacturer would not go to the expense of building an electric car capable of driving a cross-country trip, not because they could not build that car, but because the market infrastructure does not support the trip. There is simply a lack of electric battery recharging stations for the journey. Therefore, investing in building an electric car for anything other than local use would not yield an investment return at this time.

Health care has numerous examples of introduction, growth, maturity, and reinvestment. There are many great examples of how to manage the product life cycle curve.

An angioplasty balloon was designed to open blocked vessels in the heart. In the late 1980s and early 1990s, these products were in the growth phase. Revenue was increasing at double-digit rates and the average selling price was well in excess of $750. Today, according to Frost & Sullivan, the market is growing at less than 4% with an average selling price below $200.[3]

How did this happen? As the balloon angioplasty market neared maturity, a new technology, called a *heart stent*, came along to advance and grow the market. A heart stent is essentially a scaffold that is left behind to keep the vessel open. Clearly, investing today in a new angioplasty balloon would not be a prudent decision. A cycle of introduction, growth, maturity, and reinvention has kept this market growing and avoiding decline.

An example of a declining industry segment would be the dot matrix printer. Dot matrix printers are still used in certain applications, however, their use declines more every day. It would be cost prohibitive to design a differentiated dot matrix printer today at the right cost point and expect to get a return on investment. At this declining phase of the market, acquisitions and consolidation tactics would be a more profitable tactic than product introductions to make a return.

All of the above are examples of the importance of recognizing what phase of the product life cycle your market is in. A good idea is not necessarily a good product, or a profitable one, if it is not a match with its product life cycle.

Viewing the micro and macro maps and seeing what happens within a map over time helps determine the status of the industry's life cycle. If you go from one company to several companies with double-digit market growth, you are most likely moving from the introduction to the growth phase. If you go from 10 companies in a particular column to three companies with single-digit growth rates, you have probably made a shift to the maturity phase. If you notice two to three companies with low single-digit growth and recognize a product moving from one map into another, you are probably in a decline cycle.

ANALYZE YOUR ACQUIRER'S PRODUCT LIFE CYCLE

Perform the same exercise that was done with the industry product life cycle and create a potential acquirer life cycle analysis. This is equally informative. For example, if a major player does not have a product in the market or in development for a major growth segment, they are at a competitive disadvantage, which demonstrates that they may be a highly motivated buyer. If they are a weak player, it could also mean that they are about to exit the segment.

In 2017, Abbott bought St Jude for $25 billion. Looking at the concept of macro and micro grid analyses, this acquisition brought the important product categories of pacemakers and cardiac resynchronization devices to Abbott's micro map. At the macro level, these categories allowed Abbott to bring a bigger portfolio to their hospital purchasing agents who are looking to consolidate vendors and negotiate pricing discounts from a bigger base of business.[4]

Another historical example is valuable. As previously stated, Johnson & Johnson was the first company to launch a coronary stent. At the time, the company did not have the other products necessary to fill out an interventional cardiology micro map. Competitive stents were in development by companies that had a full micro map of interventional cardiology products. If one were to look at the interventional cardiology micro map in 1995, they would see that a company called *Cordis* had everything needed in the micro map. *The New York Times* reported Johnson & Johnson's acquisition of Cordis for $1.8 billion on November 7, 1995.

Now, these stories may lead the reader to believe that the world has become entirely transactional or macro map oriented; but that is not entirely true. We need to understand what each map communicates. The macro map speaks to economics, scale, and product breadth. It hints to how a company can compete in an environment that is looking for fewer suppliers and a broader base in which to negotiate price discounts. (For example, would you rather have a 10% discount off a $1 million purchasing base or a $10 million one?) However, the product themselves still need to be best-in-class and that is the story of the micro maps. The Cordis story ends in 2015 with Johnson & Johnson selling its Cordis unit to Cardinal Health for $1.9 billion. The Cordis segment requires rapid development cycles and continuous investment which was apparently not something of interest to Johnson & Johnson. As the author has no inside knowledge of the decision, we can assume either the company had better investment choices elsewhere, or they simply were slow to react to the development needs of this category. The importance of this discussion is you need to be competitive at both the macro and micro maps level simultaneously.

FINALIZE YOUR TARGET LIST AND UNCOVER THE ACQUIRER'S PURCHASE TRIGGERS

We conclude where we began: targeting requires analysis. We utilized the tools above to uncover the participants in the micro and macro markets. We looked at product life cycles of both the market and the individual companies to discern their desire and interest. Understanding the acquirer's market position allows us to identify product gaps, which are opportunities for acquisition.

From this information, the start-up should now have a list of potential acquirers and their motivations for pursuing the start-up. The next task is understanding the due diligence milestone that triggers a purchase. Investing time in pursuing an acquisition when a start-up clearly does not meet the company's purchase trigger is wasteful. More importantly, communicating before you enter the range of an acquirer's purchase trigger risks leaving the potential acquirer with an unfavorable opinion.

MAP YOUR START-UP'S EXIT POINTS

The final step of this exercise is to take your potential targets and research their acquisition histories. This can be done on a simple spreadsheet, where each row is a target (potential acquirer) and each column represents a purchasing milestone or trigger. Simple Internet inquiries will start to reveal the behaviors, practices, and personality of your targeted acquirer. When this list is done, the start-up now has a complete list of potential acquirers and their corresponding purchase triggers. The start-up has now created a map for the exit points in their industry, sorted by potential acquirer.

DETERMINE WHAT THEY WANT TO BUY: IT COULD BE MORE THAN IP

Let's take a waypoint to summarize what we know at this point and what we do not know. Our analysis of the industry via micro and macro maps, combined with understanding the industry's life cycle and the acquirer's life cycle, has allowed for an understanding of the industry's strategic future. Understanding that future, we took inventory to identify who the most likely acquiring companies are and their motivations in our investment asset (start-up). We subsequently identified each potential acquirer to understand their purchase triggers. With all of this information, we now have an *acquirer purchase trigger database* to use to align with funding milestones—more on this later.

This leaves us with one final question; what is it about our start-up that acquirers are interested in buying? Part of this answer lies in understanding our *unfair advantage*. We will expand on this concept more in our business plan section, for now, the term is used to identify the start-up's basis of competition that allows them to compete effectively with the larger companies. When most people think about the concept of unfair advantage, they immediately jump to the company's intellectual property. Clearly, this would be a predominant truth, however, there are other reasons.

In our previous discussion regarding Johnson & Johnson's acquisition of the Cordis Corporation, the motivations appeared to be the need to create a full product portfolio at the micro map level. The skilled manufacturing facility and the R&D departments may have been as valuable an asset as was the intellectual property. Perhaps the ability to immediately increase sales force presence was also important. In the case of St. Jude, there are arguable only three major cardiac rhythm patent portfolios, Medtronic, Boston Scientific, and St. Jude. St. Jude was the only company remaining as an independent. Their ultimate purchase would be inventible viewed from the perspective of the macro map.

In January 17, 2012, Gilead Sciences bought Pharmasset for close to $10 billion. It was the highest-priced pharmaceutical buyout in 2012. Gilead was interested in the company's hepatitis C drug program, as it was the most advanced program for a once-daily oral treatment.[5] In May of 2013, the company announced the product performed well in early studies and anticipated moving to a later trial.[6] On February 10, 2014, the company announced that it had submitted a new drug application to the FDA to begin these trials.[7] Without any insider knowledge, it would not be a stretch to imagine that Gilead most likely found the R&D and clinical teams of Pharmasset to be a critical part of the company's value. More importantly, all of the R&D and clinical documentation systems associated with the product's clinical development are as valuable as the IP itself. If the documentation was poorly done, the ability to gain approval from the FDA as planned is improbable regardless of the early positive results.

What were the implications of this information if you were the CEO of Pharmasset? Wouldn't it be important to make sure that you had well-organized and validated design history records for a future FDA audit? Any misstep could at worst invalidate the protocol, and at best, require costly rework and time delays. Wouldn't it be critical to have a regulatory strategy that underlies the applicant's drug development program to have some sort of FDA sponsorship? Wouldn't it be appropriate that the regulatory strategy align with the industry's future and Gilead's desires? Without FDA sponsorship, the company must validate the regulatory strategy to ensure it is viable. This could delay an acquisition. Wouldn't it also be important to have anticipated future label claims so that your trial design collects the appropriate data to support future marketing activities and labeling? What if you did not collect all the data necessary? More costly clinical trials could be required and that could be taken off the top of your acquisition price. If there are competitive programs under development, and assuming both products ultimately get approval, the stronger labeling (as proven by a clinical trial) might determine who the market leader would be. If another start-up has that better labeling, perhaps the acquirer may prefer to buy the competitive start-up. If the other product with better labeling is already in a large company, perhaps the acquirer may decide the battle for the subcategory is not worth pursuing, as it will not attain an appropriate ROI. If they are still willing to buy, the start-up's acquisition price may go down to accommodate for an ROI that the acquirer needs.

Ironically, is not this whole discussion really about predetermining which subset of the company's processes will be revealed during due diligence, which, if executed poorly, could result in a deal being delayed or not consummated? Are we not trying to uncover the other assets that may be valuable to a deal? For example, what if an acquirer were interested in a scaled manufacturing facility and during due diligence, they found the facilities in disrepair? How about the key employees? If the acquisition requires certain employees, have you motivated and secured them to stay? Of course, the opposite could be true: what if the company is only interested in your IP? Are you investing in assets that no one cares about? If so, the start-up is not going to get an appropriate return on the investment in those assets. Thus, you should be outsourcing those components. If you outsource, have you motivated your contractors and suppliers for an orderly transition?

Let's go back to our *Acquirer Trigger Database* and add some more columns to represent the questions that are appropriate to your life sciences vertical: pharmaceuticals, biotechnology tools, diagnostics, medical devices, and health care IT. Create a major heading called *Personnel* and make subcolumns for each department. For each acquirer add a "yes" or "no" to represent their interest in adding the people to their organization. If it is only a specific person or persons, identify by adding a comment to that particular box. The reason for doing this by acquirer is that each acquirer may be different. Create another heading called *Unfair Advantage* and make appropriate subcolumns. Most likely patents is a box, however, you should add the specific patents, as some companies may not be interested in all of them. Under the unfair advantage, there may also be proprietary relationships such as with a major customer or particular thought leader. Finally, add a heading called *Business Process Management*. Create the subcategories that represent the major interests of each acquirer, for example, the design history records. Returning to Chapter 9, Figure 9.1, the "Due Diligence Checklist" can provide a list of all the questions that could be considered here. Your specific analysis may uncover a question you may have not considered on the "Due Diligence Checklist" that should be added to the list.

THE ACQUIRER PURCHASE TRIGGER DATABASE

We now have a complete database that provides intelligence on the future of the industry, the motivations of our acquirers, an understanding of their timing trigger, and the other important aspects to the acquirer.

As a start-up is constantly starved and rationing cash, management now has optics to prioritize those people, assets, and processes that are important to each acquirer. As importantly, management also has a solid rationale for the areas in which they are not going to invest. Now that we know the timing of potential acquisition points, we need to align them with our funding strategy. These are the processes that could materially delay, alter, or kill a deal.

DISCUSSION QUESTIONS:

1. How is "cost of customer acquisition" derived?
2. What is an "acquirer's purchase trigger"?
3. What constitutes a "macro map" and why would you use this over a "micro map"
4. In the Product Life Cycle, what occurs during the "growth stage"?
5. Give an example of a current product and how it relates to its current product life cycle.
6. What constitutes an "unfair advantage"?

NOTES

1 History, *DePuy Synthes*, http://www.depuy.com/about-depuy/corporate-info/history.
2 5 Takeaways from Becton Dickinson's $24B Acquisition of C.R. Bard, MedCity News, Stephanie Baum, April 24, 2017, https://medcitynews.com/2017/04/5-takeaways-becton-dickinsons-24b-acquisition-c-r-bard/ (accessed February 2021).
3 Frost & Sullivan, *U.S. Angioplasty & Vascular Closure Device Markets*, F877-54, October 31, 2006.
4 Abbott-St. Jude merger gives providers what they want: fewer vendors, Modern Healthcare, Adam Rubenfire, January 6, 2017 (accessed February 2021).

5 Gilead Sciences/Pharmasset, *FiercePharma*, February 13, 2013, http://www.fiercepharma.com/special-reports/gilead-sciences-pharmasset.
6 Eric Palmer, Gilead Profits Surge 63%, But Not Where Wall Street Expected, *FiercePharma*, May 3, 2013, http://www.fierce-pharma.com/story/gilead-profits-surge-63-not-where-wall-street-expected/2013-05-03.
7 Eric Palmer, Gilead Profits Surge 63%, But Not Where Wall Street Expected, *FiercePharma*, May 3, 2010, http://www.fiercepharma.com/story/gilead-profits-surge-63-not-where-wall-street-expected/2013-05-03.

15 Align an Investor's Fundable Milestones and an Acquirer's Exit Points

Let's continue to work with our existing models to deepen our understanding of the importance of matching your investment property (also known as your start-up) with the needs of the individual funders and acquirers.

Returning to Figure 6.1 in Chapter 6, take a minute to reacquaint yourself with how start-up capital flows. In Chapter 6, we reviewed how the stages of investment flow from seed, to early stage, to growth stage, to later stage, and finally to the exit. We also discussed who the players are in start-up financing; friends and family, crowdfunding, incubators, angel capital, venture capital, and corporate venture. We discussed investor motivations and how they measure success in the companies in which they invest. In the case of corporate and venture capital, we also discussed how others measure the performance of the fund.

In Chapter 11, our *Disease State Fact Book* (*DSFB*) enabled us to understand the difference between an investment that is an incremental improvement versus one that grows the entire market. This allowed us to determine if our investment is attractive, particularly to venture and corporate capital.

KNOW YOUR TARGETS: ARRIVE EARLY AND CHEAPLY

Periodically you will hear investors speak to only investing into 10x opportunities. This concept is flawed because the issue is *not* as broad stroked as $10 million goes in and $100 million comes out. The issue is *not* about the growth rate in the valuation of the overall company. This issue is about the valuation of the stock to each individual investor and the question is how does a start-up plan for this?

The answer starts with data to create a funding map with the appropriate value and fundable milestones identified. A small tangent on the history of obtaining private equity data is helpful. Historically, most venture firms belong to an online database owned by Dow Jones called *Venture Source*. A competitive company called PitchBook was launched in 2006 and in 2020 the data assets of VentureSource were acquired by Crunchbase.[1] Both these companies track venture-backed companies in the United States, Canada, China, Israel, and India. The database can be accessed at many business schools and can be found at: www.venturesource.com. Every few years, the author downloads a subset of data

DOI: 10.1201/9780367533052-15

from one of these sources and creates a model for each life science vertical. The data offered in Figure 15.1 represents the types of companies that moved through specific classes of products from 2013 to 2018. The chart in Figure 15.1 purposely *does not* represent all source data for 2018 nor does it represent the returns of the author's funds.

The point of Figure 15.1 is that there is an independent calculation of the value milestones and the timing between milestones that represent a sample of companies. This allows companies to have a basis for planning and benchmarking.

Take a moment and look at this chart (Figure 15.1) as we define some of its content. Notice that we have all the major life science verticals represented: *Medical Devices, Pharmaceuticals, Health Care IT, Molecular Diagnostics*, and *Biotechnology Tools*. Look across the top to notice the various rounds; Series A through Series E. Let's look under the medical device segment, Series A. Notice the three boxes of *Pre-Money, Capital Raise*, and *Post-Money*—what do they mean? The premoney value is what the company is valued at before it raises its capital. The capital raised is how much funding was obtained in that round. The post-money is calculated by adding the pre-money and the capital raised. Notice in the *Capital Raise* and *Post-Money* boxes there is a +/– sign to represent the statistical ranges of the data. Continuing with the *Medical Device* category and looking under the Series A box, you will see the number of months between fund-raising activities. It took an average of 11 months +/– 2.4 months to go from seed funding to a Series A funding.

Let's go to the last column under the medical device category. One can see the details of an average medical device exit. On average, a company that made it to an exit and raised $54 +/– 15 million and exited at a value of $107 +/– $43 million. Note that the +/– represents the standard deviations in the data. An average company journey took 72 months and they achieved a 2.2 multiple and a 14.8% compounded annual growth rate.

Now that you understand the medical device details, take a look at all the verticals. Are the numbers of months between each round radically dissimilar? Look at the capital raised for each series: is it radically dissimilar? Look at the average multiples: are they radically dissimilar? The answer to all these questions is they are not radically different. Interesting, isn't it?

What is radically different is that each life science vertical has different activities that represent the value and fundable milestones for each round. As a reminder, a value milestone is the achievement of an activity that increases the value of the company. A fundable milestone is an achievement that allows the company to move to the next investor class—for example, from friends and family to angels, or from angels to venture capital. Continuing with our medical device data from Figure 15.1, in Figure 15.2 the author has placed some milestones that could represent fundable milestones. The goal of this demonstration is to represent the process that the start-up CEO should go through to create something that is specific for his industry and his product category.

Before we conclude this discussion, we need to recognize that the venture source data gives us a standardized average for our product category. Many companies will incorporate a handful of companies that exited with great success during their investment pitch. We refer to these as *comparatives*. The problem with comparatives

Dataset: 2011 Venture Source

Medical Devices

Series A — in millions			Series B — in millions			Series C — in millions			Series D — in millions			Series E — in millions			Exit Details		
Pre-Money	Capital Raise	Post-Money	Pre-Money	Capital Raise	Post-Money	Pre-Money	Capital Raise	Post-Money	Pre-Money	Capital Raise	Post-Money	Pre-Money	Capital Raise	Post-Money	Capital Raise	Exit Value	Months
2.7	1.3 +/– 1	4 +/– 2	8.0	4 +/– 1.9	12 +/– 4.3	15.9	8.9 +/– 3.5	24.8 +/– 7.6	32.1	116 +/– 5.9	43.7 +/– 15	62.0	16 +/– 5.9	78.3 +/– 24	54 +/– 15	107 +/– 43	72 +/– 21.6
11 months +/– 2.4			14 months +/– 2.5			15 months +/– 3.4			20 months +/– 8.7						Multiples	CAGR %	
															2.2 +/– 0.7	14.8 +/– 7.6	

Pharmaceutical

Series A — in millions			Series B — in millions			Series C — in millions			Series D — in millions			Series E — in millions			Exit Details		
Pre-Money	Capital Raise	Post-Money	Pre-Money	Capital Raise	Post-Money	Pre-Money	Capital Raise	Post-Money	Pre-Money	Capital Raise	Post-Money	Pre-Money	Capital Raise	Post-Money	Capital Raise	Exit Value	Months
2.8	1.0 +/– 0.6	3.8 +/– 1.6	8.0	6.7 +/– 3.4	14.7 +/– 4	21.9	12.3 +/– 6.6	34.2 +/– 10.5	51.2	20 +/– 97	1.2 +/– 22.7	89.0	25 +/– 11	114 +/– 26.6	94 +/– 38	177 +/– 65	72 +/– 20.4
11 months +/– 3			17.3 months +/– 3.4			15.7 months +/– 3.6			16 months +/– 4.2						Multiples	CAGR %	
															1.9 +/– 0.6	13.7 +/– 7.6	

Health Care IT

Series A — in millions			Series B — in millions			Series C — in millions			Series D — in millions			Series E — in millions			Exit Details		
Pre-Money	Capital Raise	Post-Money	Pre-Money	Capital Raise	Post-Money	Pre-Money	Capital Raise	Post-Money	Pre-Money	Capital Raise	Post-Money	Pre-Money	Capital Raise	Post-Money	Capital Raise	Exit Value	Months
2.0	0.7 +/– 0.4	11 +/– 2.3	5.3	3.8 +/– 1.3	9.1 +/– 2.8	14.0	5.5 +/– 2.6	19.5 +/– 6.4	23.7	5.1 +/– 2.5	28.8 +/– 13.6	23.1	4.6 +/– 2	78.3 +/– 24	19.7 +/– 9.7	48 +/– 25	60 +/– 25.2
11 months +/– 2.3			14 months +/– 2.4			14.2 months +/– 3.8			16.3 months +/– 6.5						Multiples	CAGR %	
															2.3 +/– 0.6	41.4 +/– 21	

Molecular Diagnostics

Series A — in millions			Series B — in millions			Series C — in millions			Series D — in millions			Series E — in millions			Exit Details		
Pre-Money	Capital Raise	Post-Money	Pre-Money	Capital Raise	Post-Money	Pre-Money	Capital Raise	Post-Money	Pre-Money	Capital Raise	Post-Money	Pre-Money	Capital Raise	Post-Money	Capital Raise	Exit Value	Months
1.6	0.6 +/– 1.4	2.2 +/– 1	5.9	3.7 +/– 2.2	9.6 +/– 4.3	17.0	6.1 +/– 3.3	23.1 +/– 8.7	31.5	7.5 +/– 4.6	39 +/– 15.6	21.4	13.3 +/– 6.8	34.7 +/– 17	36 +/– 14.5	94.5 +/– 41	63.6 +/– 22.8
9.8 months +/– 5			16.5 months +/– 4.3			16.9 months +/– 4.7			14.1 months +/– 4.8						Multiples	CAGR %	
															2.7 +/– 1.3	19.7 +/– 8.3	

Biotechnology Tools

Series A — in millions			Series B — in millions			Series C — in millions			Series D — in millions			Series E — in millions			Exit Details		
Pre-Money	Capital Raise	Post-Money	Pre-Money	Capital Raise	Post-Money	Pre-Money	Capital Raise	Post-Money	Pre-Money	Capital Raise	Post-Money	Pre-Money	Capital Raise	Post-Money	Capital Raise	Exit Value	Months
1.0	1.1 +/– 1.8	2.1 +/– 1.3	8.2	3.7 +/– 2	11.9 +/– 5.2	19.7	4.4 +/– 1.9	24.1 +/– 11.4	34.2	4.7 +/– 2.3	38.9 +/– 18.7	41.9	8.2 +/– 4.3	50.1 +/– 16.6	22.9 +/– 7.9	42.5 +/– 9.7	87.6 +/– 24.4
16 months +/– 6.7			17.7 months +/– 6.2			12.4 months +/– 2.7			19.1 months +/– 4.2						Multiples	CAGR %	
															1.8 +/– 0.7	14.7 +/– 7	

FIGURE 15.1 2018 market data subset PLSG analysis.

Series A — in millions		
Pre-Money	Capital Raise	Post-Money
2.7	1.3 +/− 1	4 +/− 2

11 months +/− 2.4

Fundable Milestones:

- Create proof of concept
- Hire technical team
- Demonstrate the viability of the Commercialization Plan

Series B — in millions		
Pre-Money	Capital Raise	Post-Money
8.0	4 +/− 1.9	12 +/− 4.3

14 months +/− 2.5

Fundable Milestones:

- Create prototype
- Domain mgmt, BOD, and advisor
- Define regulatory pathway and I.P. pyramid

Series C — in millions		
Pre-Money	Capital Raise	Post-Money
15.9	8.9 +/− 3.5	24.8 +/− 7.6

15 months +/− 3.4

Fundable Milestones:

- Preclinical activities
- Determine sales strategy
- Complete Mgmt Team build-out
- Patent approvals

Series D — in millions		
Pre-Money	Capital Raise	Post-Money
32.1	11.6 +/− 5.9	43.7 +/− 15

20 months +/− 8.7

Fundable Milestones:

- Launch 510k
- Clinical trials PMA and biologics
- Demonstrate sustainable high-growth revenue
- Attain positive cash flow

Series E — in millions		
Pre-Money	Capital Raise	Post-Money
62.0	16 +/− 5.9	78.3 +/− 24

Fundable Milestones:

- Launch PMA and biologics
- Scale sales team
- Determine sustainable high-growth revenue
- Attain positive cash flow

Exit Details		
Capital Raise	Exit Value	Months
54 +/− 15	107 +/− 43	72 +/− 21.6
Multiples	CAGR %	
2.2 +/− 0.7	14.8 +/− 7.6	

FIGURE 15.2 Medical device funding map sourced from Figure 15.1.

is that they represent the companies that make it through the process. Comparatives do not incorporate companies that did not make it through. The venture source data incorporates companies that did not make it through. If you are an investor and trying to incorporate the law of averages, the venture source data is best used particularly in the first few rounds. The venture source data is therefore referred to as *standard* as opposed to specific companies, which are called *comparatives*.

As we continue to look at Figure 15.2, how could we increase value to our Series A investors? The formula for return on investment contains three components, one is time, another is the investment, and the third is return. What if we were able to go from Series A post-money to the beginning of Series B in 5 months as opposed to the average of 11 months? Wouldn't that favorably impact the return on investment formula? What if the company got a $1.3 million non-dilutive SBIR grant and did not need to raise a Series A at all? The pre-money in our map for Series B is $8 million. If we offered a pre-money for the Series B of $7 million, wouldn't that be attractive to investors? Yet, because of our non-dilutive funding, we went from a Series A pre-money of 2.7 million to a Series B pre-money of $7 million. That is a 2.6× return, just between rounds ($7 million/$2.7), by receiving a non-dilutive grant.

Another example may be helpful. If the acquirer had no interest in your start-up building out its own manufacturing capabilities, might it be better to lose half a percentage point in gross margin by outsourcing manufacturing than spending $5 million in equity investment creating a manufacturing facility? If that were appropriate, wouldn't the $5 million add to the company stockholder's return on investment?

These are just examples of the kind of thinking that can occur when one has a map as to where they are going and the waypoints in between. With thoughtfulness and creativity, various tactics can be deployed at each subsequent round to bring value to each class of stockholder. The important thing for the CEO is to understand that overall valuation is not as important as the value delivered to each class of stockholder. A 10× multiple can happen in stock value and not overall valuation.

ADD THE ACQUIRER'S EXIT POINTS TO YOUR FUNDING MAP

At this point, the start-up has built the equivalent of Figure 15.2 to represent the company's own funding map with associated waypoints. The next step would be to go to your acquirer purchase trigger database from Chapter 14, and post the exit points to the appropriate Series on your map. Next, place a probability on each exit point so that you have a sense of its probability of recurrence. Most companies have two to four exit points, with varying degrees of probability.

Continuing with your funding map, the management team should pick the most conservative and most likely scenario for their exit. By staking a position on the exit scenario, the company now has a plan for how much money they will need to raise. From this vantage point, they can plot the various investor types, such as angel or venture capital, from which they will need to raise their funding.

CREATE A FUNDING RELATIONSHIP MANAGEMENT SYSTEM

Now that we have created our funding map, posted our acquirer's exit points, and identified the most likely conservative exit scenario, we can create a funding relationship management (FRM) system to support our fund-raising process. We are building a database and system that is no different than a customer relations management (CRM) system. A CRM system is a model for managing the company's interactions with current and future customers. A CRM is supported by a contact management system and integrated into the overall sales process. We are doing the same here, except we are doing it for our funding relationships and calling this our *FRM system*.

The first step is to build a supporting database by starting at the potential acquirers and working toward earlier funding. By understanding the last fundable milestone on your funding map, we can determine if your start-up is going to require corporate or venture capital. For the thoroughness of our discussion, we will assume that the last fundable milestone requires corporate or venture capital.

Now let's return to our acquirer's list. The next step would be to uncover if there are preferred syndicates of corporate or venture capital group(s) that your acquirers like to buy from. Simple searches in CB Insights, PitchBook, EDGAR, press releases, or other online sources can reveal this information. It is ironic that many start-up CEOs resist this step, complaining that it takes too long. Ironically, these same CEOs recognize that before they deploy an expensive sales force, they need to spend hundreds of hours grooming their customer target list for sales potential and invest in a supporting management system to ensure resources are managed and the organization learns through measurement. Yet, this same CEO does not see the need to manage the fund-raising process in a similar fashion.

Returning to our database and determining if there is a preferred syndicate will not change our next step. We need to start a new worksheet in the funding map spreadsheet: this one will be a sortable database. Identify the corporate or venture capitalists associated with the last exits in your preferred vertical by naming them in column A. Create a second column, column B, and title that column *Investor Type*. The columns will identify each investor type such as *Angel, CVC*, or *VC*. Create a third column, C, and title that column as *Preferred*. Place a "P," representing preferred in this column for every corporate or venture capitalist that is from a preferred syndicate. In the fourth column, column D, label the heading *Acquirer* and place the name of the acquirer in that column. In column E place the label of *Date* and enter the date of the transaction. The next six columns, F–K, should be labeled *Seed, Series A* through *Series E*. For each round in which the CVC or VC participated, place an "x" or an amount in that column. This identifies which round they participated in and how much was invested in that round. If you cannot find the amount, simply place an "x" in the column to represent that they participated in that round. Next, create a *Total* column, in column L, to summarize from *Seed* to *Series E*. Please note that some cells will either be blank or have an "x" in them. Our next column, column M, has the heading *Fund ID*. As some of the larger venture firms have multiple funds, you would place this information in this column. If they do not have multiple funds, you would leave it blank. The next column, column N, would be titled *Open/Closed*. This represents whether this particular fund is still investing. As previously mentioned, most funds invest from years 1 to 5 with the hope of having exits from years 5 to 10. After the initial database is created, we would go back to determine if a fund is opened or closed. Our next column, column O, is titled *State* and represents where the fund resides. The next column, P, is for *Investment Location*, to represent the regional distribution of where the investor funds.

We now have a database that is capable of holding a significant amount of information. From the discussions above, the start-up should now understand the total funds needed to achieve cash-flow positive and the various value and fundable milestones in between. The company should know if the acquirer has a preferred corporate or venture capital group. We should note whether the fund is opened or closed, the state in which the fund was incorporated, and a sense of the geographic locations of their investments. As with a CRM system, your FRM system should have an associated management process and a measurement.

PLOT YOUR FUNDING SYNDICATE BY WORKING BACKWARD

Now that we have a database with the venture capital firms that you initially uncovered, we can use the State data on where the venture fund is incorporated and the geographic location of their investments to extract the Angel Capital Association's data and identify the angel groups within their geographic location. When the plan is complete, we can appreciate that with each additional interaction we would continue to expand our knowledge and find new potential relationships.

CREATE YOUR TACTICAL APPROACH BY INVESTMENT CLASS

Returning back to your funding map, Figure 15.2, you have determined the entry fundable milestone for each class of investor. Now that you understand the entry point to a specific class of investor, the next task is plotting the typical funding requirements for that class of investor. For example, for how many years does that class typically put money to work? At this point, you have an overview for each class of investor and now you can determine the amounts of fund-raising that you want to be able to do with each class. This is important, particularly in dealing with venture capital funds. Venture capital funds frequently have minimum requirements for the amount of money that they are willing to invest in an individual round and in aggregate. If there is not enough room to put money to work in your company, you can accidentally be excluding that class of investor and not know it until it is too late. For example, in a company that needed to raise $70 million, they were able to raise $55 million through various angels. When the Angels were out of funding capability, there was only $15 million to put into the company. For many larger venture capital funds, that is not enough money to put to work for them. After a significant amount of time trying to raise the additional $15 million, the company was faced with a down round because they could not find sufficient bidders. As one can see with foresight provided by the funding map this situation could have been avoided. In summary, the funding map should be adjusted to account for the above situation by doing the following:

- Determine the entry fundable milestone by class
- Plot your funding syndicate exit-timing requirements
- Leave enough room for advanced investors

Once we have a completed funding map, start preparing to approach your targets. Before doing so, the following should be determined, if possible, before initiating contact:

- Determine the funder's investment objectives
- Determine the fund's timing to ensure you have time to deliver the return
- Determine management's capacity to take on another start-up

Once your targets are identified, you will use your first conversation to validate knowledge of the above. If there is not a match with your funder's targets, the start-up will gain credibility for their honesty. This credibility can be rewarded by asking the target whom they know whose funding criteria matches your investment thesis. If they are kind enough to offer some suggestions, the next request would be to ask for an introduction.

Being prepared and asking the right targeting questions is critical and saves the most valued time in the organization—the CEOs. An example is illustrative. The author was attending an Angel Capital Association meeting and was asked by a CEO to talk to a specific angel network. On the drive out to the meeting, the author had a 60-minute phone call with one of the associates who was performing the angel network's due diligence. The angel network and the CEO had been working together for a number of weeks. The associate set up a meeting for the author to meet one of the principals. At the meeting, the first question the author asked was what was their investment profile? The principal said that their investment criteria were life science companies that did not require more than $15 million of total capital. The company needed over $50 million of capital. Clearly this is not a match, yet how did the company and the angel commit so much time without asking this basic question? It is up to the start-up to create a standardized list of questions to qualify their targets.

INSTALL A MANAGEMENT AND MEASUREMENT SYSTEM

Referring back to the section on plotting the sales process, management should create a similar process for its fund-raising system. The CEO generally strategizes on the cost of acquisition for a customer but may not take the time to plot the cost of acquisition for a funder. All these measurements are very similar and should include the pipeline management report with probabilities of close and the timing of those probabilities. This allows management the ability to review and protect fundraising capabilities and timing.

Referring back to Figure 9.1, the CEO should also be reminded that this process has a feedback loop to the due diligence details that support each value in the fundable milestone.

PLAN-DO-CHECK-ACT (PDCA)

The PDCA cycle was made popular by Edward Deming, whom many consider to be the father of modern quality control. This concept is very simple. You create a plan, you execute the plan, you determine if the execution went according to plan and, if not, you act to take corrective action.

The PDCA cycle should also be applied to fund-raising. Every fund-raising call that does not go according to plan should be reviewed to determine its reason for failure. The reason for failure should result in a corrective action that should feed back into the system so that the learning can be incorporated into the next event.

DISCUSSION QUESTIONS:

1. Explain how the three components contained within the formula for return on investment interact
2. What is the purpose of plotting your funding syndicate?
3. What is a downround and what must a CEO must understand as it relates to one?
4. Explain the steps in the Funding Relationship Management system
5. Explain why it is important to create a tactical approach by investment class

NOTE

1 Our First Acquisition: CB Insights Acquires VentureSource Data from Dow Jones, CBInsights, July 15, 2020, https://www.cbinsights.com/research/team-blog/dow-jones-venturesource-valuations/ (Accessed February 2021)

16 Create an IP Pyramid for Impervious Positioning

Having a patent gives the start-up the right to exclude others from replicating its approach or product. As important as having a patent sounds, the patent's ultimate value can only be derived by understanding the market size in which the patent offers exclusivity. Life sciences is a complex business and to understand a patent's value a multilevel analysis must be completed.

The intellectual property pyramid (IPP) is an analysis technique of both the market and the patent landscape to determine if the company's IP allows them to "stake a position that is impervious to competitive advancement."[1] The start-up most likely sees their intellectual property as a basis of expressing their unfair advantage to investors. However, the ability to have a patent does not necessarily create an exclusive market position. The goal of the IPP is to understand the exclusivity it provides matched with the desired result of the approach or product. What you are trying to understand is the reason the patent exists (the problem it is solving), and then ask yourself if your patent is the only path to achieving that goal. Let's conceptualize this for a moment.

Visualize a typical organization chart with a CEO and the vice presidents that report to her. Think of the CEO as the solution needed by the marketplace and the vice presidents as the ways in which you could approach solving the problem or providing a solution. If the patent that you own were the only vice president that can provide a solution to the problem, then you would have 100% exclusivity. An extreme example might be illustrative. Picture an organizational chart where the problem that you are seeking to solve is the need to create the ability to transport people long distances over land. Immediately we think of bicycles, automobiles, trucks, and buses. Imagine for a moment that you had the exclusive right to a patent for the wheel. Clearly, this exclusivity would be a very valuable patent. However, over time perhaps jet technology becomes inexpensive and it would be deployed in the wheel market in order to obtain a piece of that market. Assuming all costs, safety, and other considerations were equal, it would not be unreasonable to expect that a subset of people would prefer the jet technology. This example points out that the wheel was a solution for a problem. The root problem remains unchanged. The wheel created and validated the market at which point in time jet technology was invented to garner a profitable piece of that market. The author continuously updates an example presentation that can be seen at: https://healthcaredata.center/commercialization-startups/develop-your-ip-pyramid/.

An exercise that the author runs with his start-up companies is to get them to focus on what the real problem is that the technology is solving. Being enamored with the exclusivity of their technical approach can inhibit technologists and marketeers from working together to create the broadest protected position.

DOI: 10.1201/9780367533052-16

Continuing our example, if the problem were adequately defined and anticipated at the beginning of the intellectual property of the wheel, and we took this technology away from its creators and asked them to solve the same problem without using the wheel, it is not inconceivable that they would have come up with other solutions. The value of this exercise for a start-up company is the ability to create the broadest intellectual property landscape conceivable. This ensures the broadest protection possible.

Let's take another example that an investor might see. The situation is that someone approaches the investor and identifies that they have a patent for the interaction of a ligand and a receptor. Receptors are proteins embedded in a cell's plasma membrane, cytoplasm, or nucleus. When a specific signaling molecule called a *ligand* attaches to the receptor, a specific biochemical cell response pathway is either activated or inhibited. This is often described as being similar to a key that unlocks a door.

For illustrative purposes, let's assume we have a patent for ligand Z, which unlocks receptor A. In our example, having a patent for the signaling ligand Z for receptor A translates to having the key to that door and seems valuable. As it is plausible that more than one ligand can unlock a receptor, the simple IPP (Figure 16.1) uncovers that ligand X also unlocks receptor A. As the function of the patent for ligand Z was to unlock receptor A, and learning about ligand X, we ultimately realize that we are not the only ones with the key to the door (Figure 16.1).

Receptor A can also activate other downstream proteins that participate in the performance of a number of different bodily functions. One can quickly become suspicious as to the overall financial impact of having a patent for ligand Z and to determine otherwise would require a significant detailed multilayered analysis.

So, what are we looking for from an analysis? We are trying to establish the marketing framework that the intellectual property is attempting to support. This connection is frequently overlooked by the start-up as evidenced by the venture capitalists saying "that seems like a technology chasing a market solution." What they are referring to is a situation where a patent has been issued but the patent does not solve a meaningful market problem. Additionally, if we are solving an important problem, our success invites others into the marketplace and our ability to hold competitors out longer has a tremendous difference on ROI. So how do we

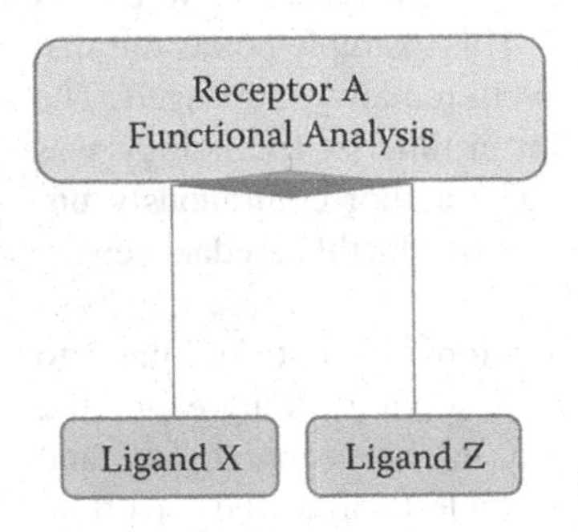

FIGURE 16.1 Ligand IP pyramid.

determine this and what filters and questions should we ask? How do we determine if the patent is solving an important market problem? The author has created a marketing framework checklist:

- Can the technology create a franchise category?
- Can the technology create a product category?
- Does the patent allow for:
  - A position that is impervious to competitive advancement
  - A clinically responsible position
  - Economic return for the user
- Is there a specific class of patient or clinical situation where the patent would be the obvious answer?
  - Is there outcomes evidence?
- How can the patent be built upon over time to evolve for distance as other players attempt to enter the market?

$$\text{Outcomes} = \text{availability} + \text{costs} + \text{quality}$$

Our first example illustrated the difficulty of using a pharmaceutical, biotechnology tool, molecular diagnostic, or health care IT patent landscapes example, as they are multilevel. Therefore let's look at a medical device for simplicity and use the procedure of balloon angioplasty as a historical example. As a reminder, balloon angioplasty is a catheter with a balloon on the end of it used to widen a blocked or narrowed vessel.

The first femoral angioplasty was performed on February 12, 1974 by Andreas Gruentzig, using a double lumen catheter. These catheters later became known as *over-the-wire catheters*.[2] Let's take a look at Figure 16.2, which is a *Disease State Fact Book* for the balloon catheter. It's important to note that this information is for illustration only, as entire books are written on the history of this market's evolution.

Using Figure 16.2, move down to row 8, which represents *Procedural Approaches*. There are three approaches available with the invention of balloon angioplasty; medical therapy, coronary artery bypass graft (CABG) surgery, and now an angioplasty procedure. Look at the circles in Figure 16.2; as they were the only mechanical solution, one can imagine that the number of CABG surgeries was likely much higher prior to this invention. The arrow on the graphic represents the probable transition that happened as historical surgical patients moved over into interventional procedures.

The over-the-wire balloon catheter intellectual property initiated the growth of the category. However, over time other methods to deliver a balloon angioplasty were developed and they too were able to receive intellectual property protection. This is represented by Figure 16.3.

Looking at Figure 16.3, we can see the pros and cons of each solution. We do not need to get into the details of each of these patents except to express the differences. For example, the monorail system had many of the benefits of the over wire system

Disease State Fact Book — Balloon Catheters

Row #			Base Year	Year 2
1	Disease Prevalence	Portion of the population found to have the condition (1 in 1,000)	24,652,555	25,268,869
2	Incidence %	Percentage of new cases (generally a year)		20%
3	Incidence	Occurrence of new cases since last time period later year or in a period of time (generally a year)		616,314
4	Percentage Recurring	Percentage of population with a recurring event in a given year		43%
5	Prevelance Population	[Disease prevalence less incidence] × percentage recurring		10,812,795
6	Number Diagnosed	Number diagnosed patients (the act of identifying treatable disease)		11,429,109
7	Diagnosis Rate %	Number diagnosed/disease prevalence (this included incident patients)		45.2%

Row #			Medical Therapy	CABG	Interventional Procedure
8	Procedural Approaches	Diagnostic, Medical Devices, Pharmaceutical, Long-Term Care, Rehabilitation, etc.	Medical Therapy	CABG	Interventional Procedure
9	Procedure/Service Approach %	The percent of diagnosed cases that would use this product/service	84.5%	3.5%	→ 12.3%
10	Number of Procedures/Services	Number of diagnosed x procedure/service approach %	9,658,740	400,019	1,406,923

11	Type of Products/Subservices	List the individual products or services performed	Over-the-Wire	Monorail	Fixed Wire	Total Balloon
12	Units per Procedure/Service	Example: 2 stents per procedure, 30 pills per cycle, 30 days in long-term care	1.1	1.1	1.3	1.18
13	Market Units/Services	Number of Procedures × Units per Procedure/Service	773,808	154,762	731,600	1,660,170
14	Average Revenue per Event	Revenue value per event or service — note revenue by manufacturer would be different than at the hospital level	$ 150.00	$ 150.00	$ 153.00	$ 151.32
15	Market Dollars or Cost	Market Units × Average Price	$ 116,071,178	$ 23,214,236	$ 111,934,824	$ 251,220,238

FIGURE 16.2 Balloon *Disease State Fact Book.* (Format from Jordan, 1996.)

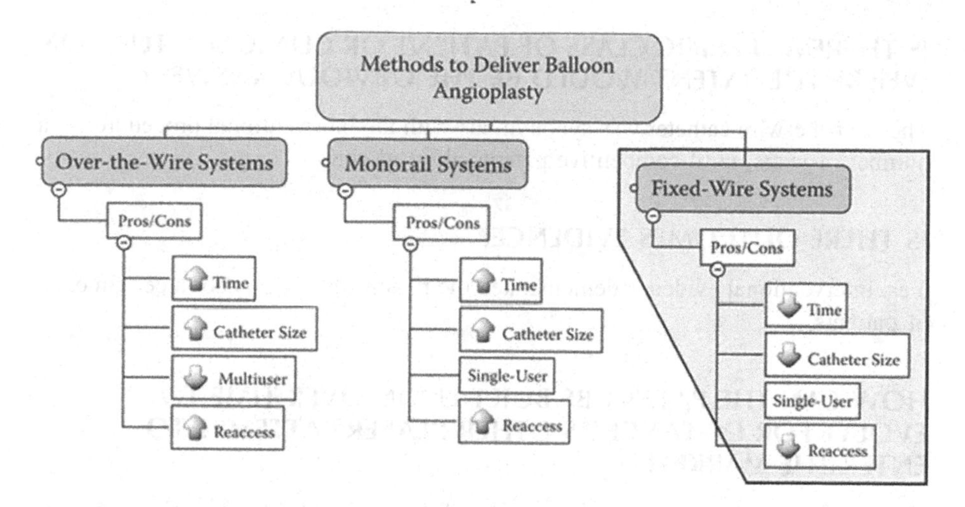

FIGURE 16.3 Balloon IP pyramid. (From IP Pyramid, Jordan, 2005.)

but allowed for a single user. The fixed wire system was faster and smaller. It benefited by allowing a single user to perform the procedure, but to its detriment, it did not have the safety profile of being able to re-access the lesion if necessary.

Returning to Figure 16.2 to the *Types of Products* line, row 11, we now see that there are three patents that have split this market. Using the IPP exercise and denying Andreas Gruentzig his own patent, could he have conceived of, and patented, a monorail and fixed wire approach? If he had, the *Number of Procedures* on row 10, Figure 16.2, could have been exclusively his for years to come.

Looking through the eyes of our marketing framework, let's review Andreas Gruentzig's IPP.

CAN THE TECHNOLOGY CREATE A FRANCHISE CATEGORY?

The over-the-wire patent offered a period of exclusivity and a short-term franchise—meaning an exclusive brand. During that time, if someone thought of an interventional approach, the over-the-wire catheter was synonymous with the approach. A long-term franchise was plausible, as the original over-the-wire patent was subsequently complimented with the monorail and fixed wire patents, resulting in a market split.

CAN THE TECHNOLOGY CREATE A PRODUCT CATEGORY?

The patent clearly created a product category.

- The patent allowed for:
    - A position that was initially impervious to competitive advancement.
    - A clinically responsible position.
- Economic return for the user over open surgery.

IS THERE A SPECIFIC CLASS OF PATIENT OR CLINICAL SITUATION WHERE THE PATENT WOULD BE THE OBVIOUS ANSWER?

The over-the-wire catheter was synonymous with the interventional procedure for a number of years, until competitive patents allowed entry.

IS THERE OUTCOMES EVIDENCE?

Yes, interventional evidence demonstrated the benefits to surgery for a certain class of patients.

HOW CAN THE PATENT BE BUILT UPON OVER TIME TO EVOLVE FOR DISTANCE AS OTHER PLAYERS ATTEMPT TO ENTER THE MARKET?

The patents were built upon by the creation of the monorail and fixed wire catheters, albeit by others. Additionally, improvements in other complimentary aspects, such as advancement in extrusion technologies and materials, benefited the category over time.

Determining a patent's value must start with an understanding as to the reason the patent exists (the problem it is solving), and then deciding the patent is the only path to achieving that purpose. The value of the IPP exercise is removing the patented solution from the toolbox of its inventors and discovering if the inventors can create other solutions. Conceived of solutions should not be hampered by economics as an approach that is expensive today may be inexpensive tomorrow.

DISCUSSION QUESTIONS:

1. What are the consequences of not adequately understanding the highest tier solution for a problem?
2. Using the DSFB for Balloon Catheters (as shown in this book), explain the relationship of how the intellectual property of such procedure initiated the growth of such category
3. Explain further how the balloon catheter technology created a product category
4. Explain further how outcomes evidence was achieved

NOTES

1 James F. Jordan, Public Presentation at Johnson & Johnson National Sales Meeting (also included in copyright 2005 TXu001308322), 2000.

2 30 Years of the Balloon Catheter—A. Gruntzig and Percutaneous Balloon Angioplasty, *Cas Lek Cesk*, 144(2):80 (2005), *Pubmed*, http://www.ncbi.nlm.nih.gov/pubmed/15730223.

Section IV

A Start-Up Must Tell a Compelling Story

17 Address Your Story to the Needs of All Constituencies

A start-up must be able to tell a compelling story. A concise message must be crafted in a way that is applicable to the individual receiver. The start-up's customer is interested in understanding how the start-up is going to solve her specific problem. Acquirers are interested in understanding how the start-up will add *unusual* returns or de-risk their business, and funders are interested in understanding their return on investment (ROI) and the associated risk.

THE CUSTOMER'S STORY

Returning to Figure 7.1 in Chapter 7, the perspective of the customer is based upon where the customer is placed in the health care systems flow chart. Regardless of where the customer is within the health care systems flow chart, he is still always trying to improve the outcomes formula below:

$$\text{Outcomes} = \uparrow \text{ availability} + \downarrow \text{ costs} + \uparrow \text{ quality}$$

Let's explore how to articulate this formula for a moment. We need to be able to understand how to discuss the formula at both the macro and micro level for the customer. At the micro level, one looks at the individual procedure cost, while at the macro level we are looking at the overall cost to treat. Using a technology that does not change procedure costs, or even increases the individual procedure cost, that allows more procedures in a fixed time increases availability and improves outcomes. A second contribution to the formula may be a simple overall cost reduction per procedure. This benefit is obvious and generally the easiest to explain. The third contribution to the formula could be an overall decrease in long-term complications, thus improving quality. Although a quality increase may have no impact on the individual procedure to the operating room, the total cost of care is certainly reduced.

THE ACQUIRER'S STORY

Without getting into a finance course, it is hard to understand how an acquirer is motivated without some basics. Whether companies are publicly traded or privately traded, they all have shareholders. In any given period of time, a calculation of earnings per share (EPS) articulates how profit is allocated to each share of stock.

DOI: 10.1201/9780367533052-17

This formula allows for comparison inside the company between time periods. In the case of publicly traded companies, it allows comparison among competitors. The formula is as follows[1]:

$$\frac{\text{Net income} - \text{dividends on preferred stock}}{\text{Outstanding shares of stock}}$$

This formula is usually calculated going back four quarters, which gives you a historical perspective of the company. This is also calculated going forward, usually in quarters, in one-year increments, and frequently in multiple years. Publicly traded companies usually provide guidance on your future earnings. When this guidance is given, forecasts become expected and that is what we will call *usual* returns.

So, the next question: how is a stock's price valued? The price of the stock is generally determined by its price earnings (PE). The PE ratio is:

$$\frac{\text{Market value per share}}{\text{Earnings per share}}$$

For example, if a share cost $15 and the EPS were $2, the PE would be $7.50. Some people refer to this as the *multiple*, as its implications are how much someone would be willing to pay for $1 of earnings. The value of this formula becomes more meaningful when comparing it against your competitor's formula. In our example, investors would be willing to pay $7.50 for one dollar of earnings. If they were willing to pay $5.00 for one dollar of earnings on a competitor's stock, clearly there would be a higher expectation for the company earning $7.50. The PE ratio discussion was included here so the reader can appreciate that it is driven by EPS.[2] Exploration of EPS causes one to appreciate that its driving force is net income.

Referring back to Figure 11.1 in Chapter 11, the *Disease State Fact Book*, the start-up is well advised to gain an understanding of how their technology could change the market. For example, factors that increase market value include heart diagnosis rate increases, procedure rate increases, and units per procedure increases. Additionally, category transitions, such as the movement from something like a bare metal stent to a drug-eluting stent in the cardiology market, also changes market value. These market factors attract acquirers because their changes to the marketplace increase the value for all those who participate. Fighting within a product category requires taking share from another. These battles are less attractive to acquirers.

The reader can now appreciate that the start-up company must work from the customer backward to earnings-per-share to understand the acquirer's motivation. Start-up companies that are generally focused on the product and its solving a customer's problem seldom take the time to appreciate the financial views of the acquirers.

So, given this discussion, what are the acquirer's motivations? Acquirers trigger an acquisition for both offensive and defensive reasons. The majority of the time, acquirers are motivated (positively) to acquire companies to deliver profits beyond *usual* returns. (This speaks to the historic and the projected returns already in place for an acquirer). When referring to an acquirer's *unusual* returns, we are referring to the desire to grow beyond existing expectations. An acquisition of a high-growth company and/or a new business segment provides this opportunity.

Sometimes an acquirer is motivated for defensive reasons, as they may buy for risk-reduction purposes. The most draconian example would be an acquirer buying intellectual property (IP) with no intention of commercializing it. This occasionally happens and it can make perfect sense if you appreciate its context. For example, if a company is a market leader and has a majority market share position, they may not be interested in shifting the market. Shifting a stable, predictable market brings risk and costs. The shift itself may change the market to unstable and unpredictable conditions, resulting in the inability to deliver *usual* returns. So why buy the technology at all? The fear would be that if a competitor got this particular IP, the market leader risks losing share or even the entire market. In multibillion-dollar markets, it may make perfect sense for the market share leader to buy a technology for $100 million and simply put it on the shelf. It is important for the start-up company to understand the acquirer's strategy and motivations for their negotiations of the exit. For example, the author has seen start-ups negotiate for lower upfront monies for the promise of a higher future royalty stream. If the acquirer has no intention of commercializing the technology, the future royalty stream is worthless.

THE FUNDER'S STORY

Life science companies have either long clinical trial pathways, as is the case for pharmaceutical and medical device companies, or lengthy market studies, as in the case of diagnostic and biotech tool companies. As such, meaningful revenue is generally 3 to 5 (or more) years out from the company's founding. As such, the early years are all about creating a compelling message to attract funders.

Until a start-up moves into the early-stage phase, they are unlikely to have deep-pocketed investors capable of funding them until they are cash-flow positive. What this means for the CEO is that a good portion of his time will be spent obtaining financing. As a result, the start-up company must understand the expectations of the funders and create an investment pitch that attracts their attention.

Please refer back to Chapter 6 where the details of angel, corporate, and venture capital were discussed. During those discussions, we discussed how to calculate different return models and, as importantly, know which investor types to use them with. As a reminder, the three predominant methods are ROI, return on multiples, and internal rate of return (IRR).

An investment pitch supports a simply stated return number for the investment property. Supporting the investment pitch is a detailed business plan, which is the subledger of all detail that supports an investor pitch.

DISCUSSION QUESTIONS:

1. Explain in further detail the customer's story
2. Explain in further detail the acquirer's story including

a. Price of stock
b. How a start-up company must work from the customer to understand the acquirer's motivation
c. What are the acquirer's motivations?

3. Explain in further detail the funder's story

NOTES

1 *Investopedia*, s.v., "Earnings per Share," http://www.investopedia.com/terms/e/eps.asp.
2 *Investopedia*, s.v., "Price-Earnings Ratio," http://www.investopedia.com/terms/p/price-earningsratio.asp (accessed March 15, 2014).

18 Deliver to Your Plan

The business plan defines the business's goals, the rationale, and evidence as to why the goals can be attained and the plan for reaching those goals and associated profits from the business plan's execution. For Fortune 500 companies, the contents of a business plan are very different than those of the start-up company. When the author has written business plans while in Fortune 500 companies, the business plan is more focused on demonstrating the ability to deliver customer revenue and corresponding profit and loss. From that basis, the Fortune 500 company prepares multiyear financial statements, which are then rolled up by each division of the company to derive the overall business plan of the corporation. Being acquired is generally not part of the consideration and obtaining additional funding, if required, is done at the corporate headquarters through the use of debt instruments or use of the public stock market. The company's historical reputation is its evidence of the ability to achieve its plans.

A start-up company does not have the benefit of a Fortune 500 company's historical reputation and evidence of its ability to perform and provide a return. The start-up company's business plan must address and articulate a compelling value to funders, acquirers, and the customer. That story must be intuitively logical and provide evidence to demonstrate that the start-up has thoroughly thought through the details of how to attain profitability in the execution of its business plan. The start-up must express why someone would acquire them to achieve their liquidity event and a rationale for its return is the predominant reason why funders are attracted.

There are multiple publicly available business plan templates that the reader can download to give guidance in the preparation of the business plan. If one searches Microsoft's Web site you will find business plan templates already in a Word format. One template that is detailed toward start-ups is the *Commonwealth of Pennsylvania's Entrepreneur's Guide: Starting and Growing a Business in Pennsylvania*. This guide can be downloaded from: https://dced.pa.gov/download/entrepreneurs-guide/.

Let's assume the reader will seek out one of these business plan templates; the author will only discuss its generic components. Instead, we shall invest our effort to detail the components and sublevel tools specific to a life sciences business plan.

MISSION STATEMENT

Your mission statement should state the purpose of your company and the marketplace in which the company intends to compete. This should be brief and allow for future growth.

DOI: 10.1201/9780367533052-18

EXECUTIVE SUMMARY

The executive summary should be an overview of the entire business plan. It should be no more than one page long and it is intended to entice the reader to want to listen to the company's investment pitch. The executive summary should be considered your 30-second commercial.

Appreciate that most corporate and venture capitalists receive hundreds of plans in a given month. Without a networked relationship into the corporate or venture capitalists, it is unlikely that your business plan would be read beyond the executive summary. As such, the business plan must convince the reader that you understand the details of the inputs, the processes, and the outputs to deliver your business plan. Referring back to Figure 8.1 in Chapter 8, your executive summary should provide confidence to the reader that you understand and can build the social network necessary for a successful exit. As important, the executive summary should highlight the company's major risks and how you plan to address them. The author reads many business plans where major risks are not addressed. Should the reader himself identify a risk while reading the business plan, the inability to acknowledge that risk and identify how it will be mitigated could be the difference between the reader reading the business plan or dismissing it with prejudice.

OVERVIEW

The overview is an introductory paragraph to your business plan expressing what the business is focused on, including your business model, and the key individuals involved in the business. It also details the legal structure.

THE OPPORTUNITY

All life science products, when taken to their conclusion, end with the delivery of health care or support the delivery of health care. As such, the opportunity must begin with the definition of the problem: the gap caused by today's existing products, and how the start-up's product is solving that problem. Once identified, the problem must be quantified by defining the market size and the share of market you expect to capture.

If the company is a diagnostic, pharmaceutical, or medical device company, the basis of the opportunity is generally disease-based. For example, if your company is a medical device company and the product participates in the interventional coronary stent procedure, how is the product changing the outcome of that procedure? If your company is a pharmaceutical company and you are in the cholesterol-lowering medication segment, you must be able to derive your benefit among the existing categories of statins, niacins, bile acid resins, fibric acid derivatives, and other cholesterol absorption inhibitors.[1] What makes you different and what meaningful improvement to outcomes would the physician (and patient) derive from your drug? With this new drug would the physician change his therapeutic protocols? If your company is in the diagnostic business and you offer blood tests for heart disease, today's tests include C-reactive protein to diagnose inflammation, fibrinogen, that determines the blood's ability to clot, and homocysteine, which your body uses to make protein and build tissues. Inflammation can insinuate risk of vessel plaque and fibrinogen or homocysteine levels can insinuate risk of stroke and peripheral artery disease. How is your test addressing an existing product gap? Would the physician therapeutically recommend a new approach to treating the disease with this diagnostic knowledge? All too often the entrepreneur misses these basic questions. Frequently scientists and technologists create start-up technology, and although the technology may perform better, its better performance may not statistically change the outcome. If outcomes are not meaningfully changed, the start-up's journey is not advised.

Under the diagnostic, pharmaceutical, or medical device scenarios where a disease or multiple diseases are addressed, the ability to summarize and articulate the opportunity by using the *Disease State Fact Book (DSFB)*, Figure 11.1 in Chapter 11, allows for the ability to pinpoint where the product benefit is derived and its associated market opportunity. The end result of the *DSFB* is market units, average selling price, and market dollars. These should be posted to your *Market Share Fact Book (MSFB)*, Figure 18.1. In scenarios where there are multiple disease states, there should be a corresponding *DSFB* and a *MSFB* for each disease state. To summarize the overall market opportunity, each *MSFB* would be rolled up into one overall *MSFB* to demonstrate the total market opportunity. Finally, by using the outcome formula below, the start-up company can connect value beyond the individual disease or procedure. For example, if the product does not change the individual procedure cost but allows for more procedures to be done in a day, this

Market Share Fact Book Date: xxx

Geography: U.S.
Disease: Cornonary Artery Disease

Startup Company, INC.
UNIT SALES & MARKET SHARE
MARKET: Stents

I. Your Company A Net Sales

	Gross Profit %	Units	Annual Inc. %	Sales	Annual Inc. %	ASP	Annual Inc. %
Yr # 1		2,217,234		$1,462,758,344		$659.72	
Yr # 2	0.680	2,408,162	8.6%	$1,456,938,105	−0.4%	$605.00	−8.3%
Yr # 3	0.695	2,833,132	17.6%	$1,679,763,933	15.3%	$592.90	−2.0%
Yr # 4	0.710	3,320,077	17.2%	$1,929,103,892	14.8%	$581.04	−2.0%
Yr # 5	0.720	3,873,423	16.7%	$2,205,608,783	14.3%	$569.42	−2.0%
Yr # 6	0.720	4,495,937	16.1%	$2,508,879,991	13.8%	$558.03	−2.0%

II. Estimated Share of Market Potential

UNITS

	Total Market		Share of Market						
	Units	Annual Inc. %	Company A	Company B	Company C	Company D	Company E	All Others	Total
Yr # 1	2,463,593		90.0%	5.0%	5.0%	0.0%	0.0%	0.0%	100.0%
Yr # 2	2,833,132	13.00%	85.0%	5.0%	5.0%	5.0%	0.0%	0.0%	100.0%
Yr # 3	3,541,415	25.00%	80.0%	6.0%	6.0%	5.0%	3.0%	0.0%	100.0%
Yr # 4	4,426,769	25.00%	75.0%	9.0%	7.0%	5.0%	3.0%	1.0%	100.0%
Yr # 5	5,533,461	25.00%	70.0%	10.0%	8.0%	6.0%	3.0%	1.0%	98.0%
Yr # 6	6,916,826	25.00%	65.0%	14.0%	9.0%	7.0%	4.0%	1.0%	100.0%

Dollars

		Total Market		Share of Market						
Mkt ASP		Dollars (000)	Annual Inc. %	Company A	Company B	Company C	Company D	Company E	All Others	Total
$625	Yr # 1	$1,539,745,625		95.0%	5.0%	0.0%	0.0%	0.0%	0.0%	100.0%
$550	Yr # 2	$1,558,222,573	15.00%	93.5%	4.0%	2.5%	0.0%	0.0%	0.0%	100.0%
$539	Yr # 3	$1,908,822,651	22.50%	88.0%	5.0%	5.0%	2.0%	0.0%	0.0%	100.0%
$528	Yr # 4	$2,338,307,748	22.50%	82.5%	7.0%	6.0%	2.0%	1.5%	1.0%	100.0%
$518	Yr # 5	$2,864,426,991	22.50%	77.0%	9.0%	7.0%	4.0%	2.0%	1.0%	100.0%
$507	Yr # 6	$3,508,923,064	22.50%	71.5%	12.0%	8.0%	5.5%	2.0%	1.0%	100.0%

FIGURE 18.1 *Market Share Fact Book.* (From Jordan, 1991.)

improves availability. We are reminded that, in the end, the life sciences start-up company must be able to also articulate the value at the macro level. Did the opportunity demonstrate an outcome?

$$\text{Outcomes} = \uparrow \text{ availability} + \downarrow \text{ costs} + \uparrow \text{ quality}$$

If the company is a biotechnology tool company or a health care information technology company, the ability to pinpoint where the technology will be applied and who is serviced in the health care systems flow chart (HCSF), Figure 7.1 in Chapter 7 helps articulate the opportunity. For example, in the biotechnology tools segment, a company that offers a cellular systems biology solution could allow for more accurate predictions of drug safety and efficacy. Utilizing Figure 7.1 in Chapter 7, the tool can articulate where in the health care value chain this opportunity exists. Utilizing Figure 1.2 in Chapter 1, and its highlighted dotted box, further narrows where this biotechnology tool benefits the drug manufacturer's drug development value chain. Next, the company would need to forecast the companies and the number of clinical trials to anticipate which companies could benefit by the use of the start-up's technology. To demonstrate the company's knowledge of their market, they may need to provide a macro or micro map of the companies that could benefit from their product—see Figures 14.1 and 14.2 in Chapter 14. Next, the company would demonstrate that they know when the individual clinical trials would start and the corresponding number of patients they would expect in each trial, see Figure 18.2. All of these documents would again feed into the *MSFB*, Figure 18.1, to forecast their market share, price, and profitability.

If the company is a health care information technology company, utilizing the HCSF, Figure 7.1 in Chapter 7 allows for the identification of the location of the

Entry Schedule Fact Book Date

Competitive Entry Schedule or Drug Trial

Company	Competitive Platform	Year # 1				Year # 2				Year # 3				Year # 4			
		Q1	Q2	Q3	Q4	Q1	Q2	Q3	Q4	Q1	Q2	Q3	Q4	Q1	Q2	Q3	Q4
Company 1	Product A																
	Product B																
Company 2	Product C																
	Product D																
Company 3	Product E																
	Product F																
Company 4	Product G																
	Product H																
Company 5	Product I																
	Product J																

Legend:
C = Clinical Trial Begins
D = Regulatory Submission
R = Regulatory Approval
L = Launch

FIGURE 18.2 Entry Schedule Fact Book.

services to be provided. For example, if the benefit were derived at the hospital, the American Hospital Association would provide the evidentiary data that there are 5,723 hospitals in the United States in 2012.[2] However, the start-up would need to deepen their analysis to generate a subledger for that opportunity. For example, how many hospitals would be interested in the company's products? Who are the competitors that exist in that space? What problem are they solving? How was the start-up solving the problem and do they have the ability to access the sales channel? Again, many of the tools provided in this book can be utilized to provide your supporting documents. As stated before, if detail cannot be easily articulated and translated into the outcomes formula below, the business plan may prove to be ineffective.

$$\text{Outcomes} = \uparrow \text{ availability} + \downarrow \text{ costs} + \uparrow \text{ quality}$$

THE MARKET

As was detailed in the opportunity section above, a start-up company must be specific about identifying their markets. Most companies require multiple products to achieve their full potential. For example, a surgical robotics technology may have an initial target market focused on throat surgery, however, the entire general surgery market may be their true long-term opportunity. Detailing the full market opportunity and the tactical approach to capturing the various market components provides great creditability to the start-up. The company can use the *Entry Schedule Fact Book* (*ESFB*), Figure 18.2, to articulate its timelines.

In detailing the start-up's tactical approach, it is helpful to provide your rationale for doing so. For example, is your rationale for your approach due to ease of customer access? Is it due to an underserved customer segment and therefore will there be a limited competition? Is it due to a regulatory risk reduction strategy? A tangential example may be illustrative. In one start-up company with which the author is associated, the biggest market is not the first target. The rationale for this approach is that the regulatory, reimbursement, and sales channel access risks are lower. The market provides for more rapid attainment of cash-flow positive, and once achieved, provides for an early acquisition trigger point for an acquirer. If the company is not acquired, demonstration of market acceptance and management's ability to get to market would make raising capital easier and less dilutive to existing stockholders.

Returning from our tangent, the market section also requires a brief discussion on intellectual property, regulatory, and reimbursement strategy. As demonstrated in our tangent above, initial market selection is frequently based upon this information and may not necessarily be the biggest market opportunity. To reiterate a previous point, if the market selection criteria are not provided, the reader is left to interpret the rationale based upon his or her personal bias. This may not be favorable to the company.

Before we conclude our discussion, it is important to recognize that start-up products enter markets that are at specific phases of their life cycles. Revisiting Figure 10.1 in Chapter 10, the product life cycle curve, one risk for a start-up is launching into an unprepared market. For example, prior to health care reform, a start-up focused on health care information exchanges would have been poorly received. The reason would be that a large supporting market infrastructure would be necessary to support the idea of an exchange. Prior to health care reform, only a large company with vast resources could approach such a monumental task. Today, new government regulation supports the creation of the health care exchange and the government is funding the necessary components of a health care information exchange. A start-up with a supporting product could potentially be successful. Start-up companies need to be aware that if an investor had a historical failure in a HIT exchange company investment, they will readily dismiss your opportunity without an education that their previous failure was due to implementing a good idea at the wrong time in a product's life cycle. Without this rationale, historical experience would cause an immediate dismissal of this opportunity.

THE COMPETITION

This section should express who is in the marketplace today, the position you will play in the market, and how you are going to win. This is a general overview, however, it should be supported by the use of macro and micro maps, Figures 14.1 and 14.2 in Chapter 14, respectively. These tools express how firms participate and compete in the marketplace and they are valuable for several reasons.

One value is that they provide an overview of the competitors in the market. They quickly articulate what segment each competitor competes in and the numerous product categories they provide. It is very difficult in a rapidly changing environment for investors to keep up to date on all the competitive landscape changes that occur daily. In the absence of expressing the entire competitive landscape, the start-up is leaving the context of the market to the reader. The author looks at these markets daily and at least once a month becomes aware of an industry acquisition that occurred or a change in market leadership of which he was unaware. The inability of the business plan to bring the reader up to date risks the plan being dismissed based upon a historical reference by the reader.

Another benefit to mapping is the ability to articulate your acquirer strategies. The micro and macro maps offer clear evidence as to who, and who is not, participating in a particular market segment.

The micro maps provide details on the individual products in a category. For example, Figure 14.1 in Chapter 14 provides a historical micro map on the cardiac rhythm management market space. In this historical perspective, the map is after Boston Scientific bought Guidant, Inc. The micro map is valuable in determining how a start-up company would compete in this market. This analysis could be further supported with an *ESFB*, Figure 18.2, that shows who is currently predicted to enter the market in the coming years. The combination of the micro map and the *ESFB* would clearly demonstrate one's knowledge of the micro market and be one basis for an acquisition discussion.

However, the micro market participants may not be the only companies interested in acquiring the start-up. The macro map, Figure 14.2 in Chapter 14, shows how the category of cardiac rhythm management fits within the various hospital provider areas. The provider areas are intersected with the companies that sell products into those provider areas. Increasingly, hospitals are responding to price pressures by establishing formulary committees to review and select the products to be used within the health care organization. Although health care product manufacturers may not be fond of formularies because they set price and limit selection, they can still use them to their advantage. At the macro level, large companies can use a formulary to create a competitive advantage. For example, if the company has a weak micro product, they could maximize its use by bringing it into a formulary via their macro influence. How this translates into real life is that a company could say something like, we will give you a 1% discount on our operating room products, if you buy our interventional cardiology product at some negotiated volume. (It should be noted that this works when the competitive products are incrementally better. It does not work if the product is statistically better.)

For large corporations having a full portfolio of products at the macro level, a micro product could become important to their overall macro strategy. For example, prior to Guidant being purchased by Boston Scientific, neither Johnson & Johnson nor Boston Scientific had a meaningful micro cardiac rhythm management (CRM) portfolio. In the case of Johnson & Johnson, they could negotiate at the macro level because of their significance in the operating room. Johnson & Johnson's operating room presence was so significant that they could negotiate against Guidant and Medtronic, who would strategically play the CRM card at the macro level. Unfortunately at that time, Boston Scientific could not negotiate as favorably in either the operating room or the CRM segment. Although the author has no inside information as to why Boston Scientific bought Guidant, Inc., the macro and micro maps provide for one logical reason as to why.

In our example above, we see the value of the macro map, the micro map, and its supporting *ESFB* in detailing the market. Our formulary discussion expressed the value of the tool and the various motivations of an acquirer. However, our discussion on the formulary also expresses the challenge of the start-up entering the sales channel. A formulary can block a start-up to the benefit of the incumbent if the start-up product's outcome is incremental. However, a formulary strategy cannot block a product that offers meaningful improvement in outcomes.

THE UNFAIR ADVANTAGE

In the start-up world, the concept of unfair advantage is different than the legal term used by the SEC for unfair advantage. The legal term speaks to the illegal tactics deployed by firms to exclude competition. This is not the same definition used in the start-up, angel, corporate, and venture capital worlds. In fact, the term is used to identify the start-up's basis of competition that allows them to compete effectively with larger companies. When we use the term *unfair advantage* in this book, this is what we are referring to.

PROPRIETARY RELATIONSHIPS

Return for a moment to Figure 8.1 in Chapter 8, where the social network for the start-up is described. This is a great way to start our discussions, as unfair advantage can be about proprietary relationships such as:

- A proprietary relationship with capital providers
- A proprietary relationship with customers
- A proprietary relationship with acquirers
- A proprietary relationship with channel partners and suppliers
- A proprietary relationship with a government agency
- A proprietary relationship with regulators
- Famous and powerful people

For example, if the start-up has a novel idea and a super angel commits to funding the entire idea, it is clearly de-risked compared to others who do not have such sponsorship. Should a start-up company have access due to a proprietary relationship with a major hospital chain, customer access, and market acceptance is de-risked compared to others in their investment class. Should the start-up have a significant relationship with an acquirer, the opportunity is de-risked as acquirers like to buy companies from those they know. The ability to have a relationship and gain access to a major channel partner or supplier can be very attractive to investors. For example, many start-up companies must prove their value by generating $1 million to $2.5 million in revenue before accessing a nationally ranked distributor. However, should the start-up be given access because of a prior relationship, this reduces the cost and time needed by a traditional start-up, which must spend millions to demonstrate its value to gain access to a national distributor. Many life science start-ups benefit by having thought leaders that have relationships with government grant authorities. These branded thought leaders have a greater probability of delivering non-dilutive funding to the company than if the grant was submitted by someone unknown to the authority. Individuals with relationships to regulators may have the capability to have access to information regarding regulatory trends or upcoming requirements that is not generally known to the public, or the reputation of an individual with a regulatory organization may provide a greater confidence to the FDA. This confidence may translate into the start-up company having their protocol accepted because of the reputational rigor of the known individual. It intuitively makes sense to spend more oversight on someone who is unknown. Last, people that are known tend to have greater access: surely a relationship with Bill Clinton would open doors.

PROPRIETARY KNOWLEDGE

Proprietary knowledge describes a scenario where an individual possesses technical knowledge of a product or the inner workings of a process. This individual can have personal authority by being a known thought leader. For example, someone who had invented a previous technology and had commercial success has credibility when stating they have a new idea of technical merit. A scientist who has received the Nobel Prize in a startup's technology area certainly would instill confidence in investors. An individual may also have experience in the proprietary know-how that is not patented and may have been acquired through an apprenticeship scenario. Early in the author's career in the defense industry, he met an individual who could slice wafer fabrications at a yield that was 20% better than anybody else. At the time, this talent could not be automated and every few years or so this individual would be financially incentivized to leave the organization for a start-up. A former executive from a specific industry can de-risk an opportunity by possessing inside information (not the illegal kind) of an organization or its industry. Should a start-up have this individual in their start-up, one would expect that opportunity to be de-risked.

INTELLECTUAL PROPERTY

Intellectual property includes patents, trademarks, and copyrights. These are legal rights given to their owners for exclusive use. According to the U.S. Patent and Trademark Office (USPTO) Web site, "a trademark is a word, phrase, symbol or design, or combination of words, phrases, symbols or designs, that identifies and distinguishes the source of the goods of one party from that of another." A service mark is the same as a trademark except it identifies the source of the service rather than an individual product. A copyright is different than a trademark in that it protects rights to an original artistic or literary work. For example, the movie *Superman* may be copyright protected, while the Superman logo on a T-shirt is trademarked.[3]

One of the biggest missed opportunities in start-ups is, when a new category is created, to not spend the time to develop a trademark or trade name that defines the category. Frequently, due to long regulatory and clinical pathways, the first company into a new category may have 12 to 18 months of product exclusivity before competitors follow into the category. There is an opportunity for these companies to develop a name for their product that also defines the category. With this foresight, they can brand the category to their favor and inhibit others from using their name. For example, if someone owned the name of stent, anyone performing a similar procedure would need to use another name. Think about the name Kleenex, which is simply facial tissue. That name has become synonymous with the category and as a result garners more sales than it would if it was simply just facial tissue. In the pharmaceutical industry, the name Lipitor has such recognition that even years after the category has gone into a generic status, the number of Lipitor prescriptions filled are still expected to be $3 billion a year in 2015.[4] This is the value of creating a brand.

By far the most favored unfair advantage is a patent. A patent is a property right granted by a government. In the United States, the USPTO defines a patent as a right of an inventor "to exclude others from making, using, offering for sale, or selling the invention throughout the United States or importing the invention into the United States." Patents are not just a U.S. right, most countries issue patents under country-specific rules. Due to the importance of this category, a start-up company must have legal resources dedicated to the effort.

A patent may or may not guarantee marketing and sales success. Referring back to our intellectual property pyramid exercise in Chapter 16, the goal of a patent is to create a protectable market position. A start-up company should attempt to utilize their patents to stake a "competitive position that's impervious to competitive advancement."[5]

A great example of protecting a category is how Johnson & Johnson licensed, created, and launched the stent category for use in peripheral arteries and for coronary arteries in 1991 and 1994, respectively. The company quickly captured 90% of the market and subsequently bought all patent rights from Palmaz, Schatz, and Ramano in 1998. As strong as the intellectual property was, they could not inhibit competitive entry, however, after 12 years of litigation, historic damages in favor of J&J were awarded.[6] Although Johnson & Johnson could not keep others from entering the market, their intellectual property strategy transitioned the stent design debates into two market categories; closed cell stents and open cell stents. Johnson & Johnson's patents allowed them to create the closed cell stent category for which IP protection is still valuable to this day.

The story speaks to the value of utilizing intellectual property to create market categories. Without getting into a debate on which stent design is better, open cells or closed cells, Johnson & Johnson's Cordis franchise created two categories, one for them (closed cells) and one for everybody else (open cells). Although Cordis critics will argue about the shifts in revenue, management changes, and other performance issues in the Cordis franchise over the past 20 years, one has to recognize that this one strategic and thoughtful move has allowed the company to stake a protectable franchise since 1991.

In fact, the Johnson & Johnson story expresses why funders favor patents over all other intellectual property. Patent protection can de-risk an organization that may occasionally stumble. For a start-up without that patent protection, the stumble may be deadly. For a start-up with meaningful patent protection and aligned marketing positioning, the stumble can be recovered from.

The goal in this section of the business plan is to be able to articulate one's patent position concisely. Hopefully, the start-up can also align the patent with its marketing position so it can express how it is going to "stake a position that is impervious to competitive advancement."[7]

REGULATORY AND REIMBURSEMENT

It is most likely that a discussion of regulatory and reimbursement was included in the market section of the business plan. The start-up company may or may not choose to include this section, based upon existing regulatory and reimbursement conditions. If the regulatory and reimbursement issues are stable in the start-up's particular marketplace, then this section may not be necessary. For example, if the start-up company is following in the footsteps of a well-worn pathway frequently trod by Fortune 500 companies, this may be all the evidence that is needed.

However, at the time of this writing, there are unstable market conditions caused by debates to changing regulatory and reimbursement standards. For example, in the case of medical devices, the medical device classifications and exemptions categories of the 510k, the 510k with an investigational device exemption (IDE) and pre-market approval (PMA) are all under debate. In the author's experience, the difference between 510k and a PMA could be $50 million or more. If the start-up makes an uninformed choice, this can be debilitating. In the pharmaceutical category, cellular drug therapies are holding the promise of carrying out functions that cannot be achieved with traditional drugs. However, the testing standards for this class are evolving, and the FDA, with industry, continues to learn how to manage this emerging class drug. From a start-up perspective, learning means an evolving process, which equates to risk. In the diagnostic industry, there is debate if the FDA should increase regulation beyond today's Clinical Laboratory Improvement Amendment (CLIA) standards. For individuals investing in this category, uncertainty equals risk.

In the area of reimbursement, existing reimbursement codes continue to decline year after year and the ability to obtain a new code becomes more challenging. There is probably no more challenging area in the diagnostic category. On the one hand, a CLIA status allows the company to market its product earlier than the pharmaceutical or medical device class of products. However, each diagnostic must obtain reimbursement: to do so frequently requires a clinical trial and a new code must be obtained from Centers for Medicare & Medicaid Services (CMS) in order to scale revenue. This can take several years and tens of millions of dollars.

Corporate and venture capitalists tend to invest in specific industries and are very well aware of these changes and risks. In the absence of discussing the rationale behind your start-up's plan, the reader of your business plan will judge the viability of your selection based upon their historical bias. Previously in Chapter 11, we spoke about the CEOs who presented to a venture capitalist. The first CEO was second-guessed on every regulatory, intellectual property, and reimbursement strategy that he presented. The second CEO had hired the best in class firms in each area. When he expressed his strategies followed by the branded reputation of the nationally known firms, the venture capitalist's biases were silenced by the confidence of the CEO's team. In this section of the business plan, the start-up is well advised to provide such evidence, to remove the bias of the reader, and demonstrate confidence through the independent validation point offered by a branded third party.

THE COMPANY AND MANAGEMENT

This section includes a brief history of the company and its founders. The CEO and his management board's qualifications are discussed along with that of the board of directors and the advisory board.

Referring back to a social network, Figure 8.1 in Chapter 8, the goal of this section is to communicate how the management team's skills de-risk various aspects of the company and how their social network allows the company to gain access to the various areas covered in Figure 8.1.

THE COMMERCIALIZATION PLAN

The commercialization plan provides the operational and tactical details of how the organization intends to deliver on the business plan and generate its associated revenue, profits, and generate an exit or liquidity event.

The commercialization plan communicates the efforts completed to date and maps out the details required by the start-up to move forward. The commercialization plan communicates the major product development, clinical and regulatory activities, along with scaling investments in operations, logistics, and so on, necessary to achieve the plan. These activities are quantified so that period expenses can be defined.

CAPITALIZATION AND EXIT STRATEGY

This section could be considered an extension of the commercialization plan. The commercialization plan's period expenses are matched to value and fundable milestones, and in many cases, the commercialization plan must be adjusted to align with value and fundable milestones.

This back and forth analysis starts to define and create the company's fund-raising strategy. In addition to fund-raising milestones, the start-up company should also predict the various exit points. Typically, a start-up company has more than one exit point. For example, the company could be bought when their patents are issued. They could be bought when they obtain specific regulatory approvals, or they could be bought upon a certain demonstration of revenue.

Commercialization plans that distinguish themselves recognize that all activities do not go as planned. As such, commercial plans that create contingency plans for different scenarios generally are deemed more credible. Aligning with the company's commercialization plan, financing, and exit strategies is critical. Start-ups that have various scenarios and align those scenarios are deemed more sophisticated than those that only envision one path. The challenge in the business plan is to demonstrate that the various pathways all lead to an acceptable ROI or multiple for the investor.

FINANCIAL PROJECTIONS AND RISKS

The section should summarize the major milestones from the commercialization plan. It should articulate the anticipated funding milestones and exit points from the capitalization and exit strategy section of the business plan. These milestones are then converted into 5-year projections for the company's income statement, balance sheet, and statement of cash flows. The author assumes that the reader has some experience in creating financial plans. However, if the reader is not familiar with how to create a plan, a nonprofit organization of retired executives called *SCORE* provides templates that can be found at: http://www.score.org/resources/business-planning-financial-statements-template-gallery.

Assumptions used in the preparation of the financial projections should be detailed and discussed. It would also be valuable to benchmark how the start-up's financial projections are aligned with those of their targeted acquirers. One of the biggest mistakes in financial projections that the author sees is in the area creating the company's balance sheet. Companies appear to be able to project revenues and associated expenses but find it difficult to project the assets that they will need to run their business. Benchmarking the sales-to-asset ratio of your potential acquirers allows one to benchmark the ratio of assets necessary to generate a dollar of revenue. The start-up can then use this to benchmark where they should be. Utilizing Hoovers online database, which can be accessed in most libraries, allows for easy access to this information. Another example of the importance of benchmarking is in the area of gross margins and net income. If an acquiring company generally has a 60% gross margin, it would be important in this section to identify whether or not the company is aligned with that strategy. If the company were not aligned with that benchmark, it would be important to express if it ultimately intends to match that benchmark. It is also important to benchmark the company to other start-ups in the category. For example, MD&DI, an industry trade association, has published that a medical device 510k generally takes $24 million to get to market. If you are a 510k and are expressing that it takes you $5 million to get to market, you most likely have something special in your start-up that mitigates that spending that should be articulated in your plan, or you are naïve and missing many of the necessary commercialization components. Local incubators, the online database Venture Source, and others provide an objective benchmark. If you are dealing with corporate and venture capitalists, they most likely know their benchmarks. If your business plan does not align with their expectations, your business plan will be dismissed with bias. In order to avoid this, the standard must be recognized in the business plan and a discussion to reconcile your plan to the benchmark could avoid being dismissed due to bias.

Finally, your business plan will continue to evolve over time and will most likely be updated with each investor round. It is important to note that your business plan, when routed to investors, becomes a public document. Future investors will most likely use your historical business plans to judge the organization's ability to create and deliver on a plan. An organization that does not consistently deliver to plan is deemed more risky than one that consistently delivers.

The end result of all this detail is the ability to determine a total amount of capital that is needed and the ability to forecast exit points and exit amounts to derive the projected returns to an investor. Fundamentally, this final calculation is used as one of the first slides in your investor pitch.

DISCUSSION EXERCISE:

1. For students developing a startup, please discuss your pitch with the class and get feedback.
2. It is important to recognize that start-up products enter markets that are at specific phases of their life cycles. Please expound
3. What are the benefits of mapping? Explain each benefit in detail
4. Many life science start-ups benefit by having thought leaders that have relationships with government grant authorities. Please expound.

5. What is the value of creating a brand?
6. Explain the importance of benchmarking

NOTES

1 High Cholesterol: Cholesterol-Lowering Medications, *WebMd*, http://www.webmd.com/cholesterol-management/guide/choles-terol-lowering-medication.
2 Fast Facts on U.S. Hospitals, *American Hospital Association*, http://www.aha.org/research/rc/stat-studies/fast-facts.shtml (accessed March 15, 2014).
3 Trademark, Copyright or Patent? *United States Patent and Trademark Office*, http://www.uspto.gov/trademarks/basics/trade_defin.jsp.
4 Linda A. Johnson (Associated Press), Against Odds, Lipitor Became World's Top Seller, *USA Today*, December 28, 2011, http://usatoday30.usatoday.com/news/health/medical/health/medical/treatments/story/2011-12-28/Against-odds-Lipitor-became-worlds-top-seller/52250720/1.
5 James F. Jordan, Public Presentation at Johnson & Johnson National Sales Meeting (also Included in Copyright 2005 TXu001308322), 2000.
6 *Wikipedia*, s.v., "Julio Palmaz," http://en.wikipedia.org/wiki/Julio_Palmaz (accessed March 15, 2014).
7 James F. Jordan, Public Presentation at Johnson & Johnson National Sales Meeting (also included in copyright 2005 TXu001308322), 2000.

19 Tell a Compelling Story with the Investor Pitch

When the author first entered marketing, his vice president constantly drilled into his organization the "rule of three." The "rule of three" was that, when selling, the marketing or sales professional had 30 seconds to entice a customer's interest. If that was successful, they might get 3 minutes to expand on the concept, and if they were highly successful, they could advance to a 30-minute discussion. This rule recognized the necessity of creating a hierarchy of messaging that aligned with the time allotted.

The author has adjusted this sales concept to the startup world where the "rule of three" is adjusted to 3 minutes, 30 minutes, and 3 hours. In that spirit, the start-up company must build a hierarchy of disciplined messaging that matches the occasion. A typical investor pitch is 30 minutes long, with approximately 15 to 20 minutes of presentation and 10 to 15 minutes of Q&A. Should this presentation be successful, a subsequent meeting of approximately 3 hours will most likely be requested to go through the additional details of the business. It's important to point out that this is where the value of a preconceived due diligence process avails itself—see Figure 9.1 in Chapter 9. A due diligence process that is aligned with the company's marketing messages, and, specifically the "rule of three," will quickly impress.

The following headings outline an investor pitch and we will discuss the details behind each component.

THE PREVIEW

The preview is the first slide and it should concisely express the opportunity in two to three sentences. For example:

> NewCo is entering the $1 billion "name market" with a product that solves the problem of "x" as evidenced by "scope the problem in dollars and in terms of the outcomes formula." Newco will require "x amount in funding" and deliver "$x" in revenue in "x" years. Newco will exit with "x years" and provide an exit multiple of "x" and an IRR of "x."

> Without disclosing confidential details, the author's incubator has invested in a robotics start-up. Internally, we create an Investment Memo for our investment committee as to why we should invest in a specific company. We created the following to articulate this particular investment:

> The major inhibitor to reigniting the double-digit growth of the $5B minimally invasive surgery market is the constraints of conventional straight tubes in reaching

DOI: 10.1201/9780367533052-19

anatomy. NewCo has developed a patent protected, flexible robotic platform that unlocks access difficulties. In doing so, the company will raise "$x," achieve "$x" revenue in 5 years, and provide an exit multiple of "x" by "name year."

As you can see, the preview requires you to articulate your value very quickly and could arguably be considered your 30-second pitch.

THE VENTURE CONCEPT

The intention here is to expand beyond your preview and spend up to 3 minutes articulating why an investor should invest in your start-up. It is critical that you recognize that your message is specific to the individual class of investor. The goal of the concept is to also handle early objections: if you cannot anticipate and handle objections upfront, they will consume the listener's mind until it is relieved.

For example, in the case of the robotics company discussed above, a large amount of capital is necessary to deliver the technology. An angel investor may fear that the product does not work and that they would get diluted as the concept is proven. How would you anticipate this and handle the objection early? If you knew the technology was previously proven and used in 911 search and rescue, would that alleviate some concern? In another example, a local drug company had received a large amount of non-dilutive funding, which significantly de-risked their investment compared to other pharmaceutical investments. Unlike other investment classes, such as medical devices or health care information technologies, pharmaceutical companies tend to fail or achieve large returns. One can imagine that many angel investors would be fearful of this class of product. How would you handle that objection? Once the start-up company anticipates the objections, they can easily incorporate its counterbalance in its 3-minute pitch.

Inclusive in the venture concept is a brief discussion of past funding and anticipated future funding. This is followed by a brief description of the strength of the management team, which includes the management board, the board of directors, and the advisors. This is not the time to go through every member, it is the moment to utilize your top members to express credibility and create comfort.

Think for a moment what we have done here in a matter of minutes. We have built upon and reinforced our 30-second pitch with more detail. We have anticipated major objections and offered a counterbalance. We have articulated past funding as a surrogate of worthiness and forecasted future funding to provide perspective. We have highlighted key members of the management team whose intention is to bring comfort and familiarity.

Although there are facts, figures, and projections in our discussion, one can start to appreciate that we are trying to anticipate emotions and prepare the listener to receive the rest of our message.

Below are the contents of the PowerPoint slide:

- 3-minute pitch
- Funding: past and future
- Strength of the team

THE MARKET NEED

In this section, the company must articulate that they have identified a clear need or problem in the market. By using the outcomes formula, they must articulate the customer's pain in both clinical and economic terms.

Once the need is clarified, a discussion as to how that need is addressed today should be at hand. Next, a discussion on the gap that exists today between the need and the available solutions should be provided. For example, prior to the availability of nonsurgical AAA stent grafts in the treatment of abdominal aneurysms, surgery was required. Surgery was the solution and the gap was that the surgery had an average mortality rate of 3%[1]–6%.[2] Unfortunately, a majority of the patients were elderly and they could not tolerate the surgery for other health reasons.

Let's look at our toolbox to see if there is something that can help articulate this problem and gap. Returning to Figure 11.1, the high mortality rate would translate into a much higher *Diagnosis Rate*, line 7, than *Procedures Rate*, line 9. The difference between line 7 and line 9 would be the gap and, if memory serves, in 1996 that gap was 10:1. In cases such as this where the gap is so clear, the author sometimes substitutes Figure 11.1 in Chapter 11 in place of a more traditional PowerPoint slide.

The last part of this discussion is one of the primary and secondary customers associated with this clinical problem. For example, an insurance company would be a secondary customer in this example. As an aneurysm may take years to grow, some insurance companies might be motivated not to treat in the hope that the patient would be in a new plan by the time the aneurysm was urgent. The start-up company should ponder how it would handle this objection. Is it statistically meaningful in the execution of their plan? If not, the existence of the situation is not material; if it is, the company must have an action plan to overcome the obstacle. The importance of this discussion is in realizing that the presenter must recognize secondary customers and their potential obstacles and have an answer should a question be asked.

Below are the contents of the PowerPoint slide:

- Has the company identified a clear need/problem in the market?
- The clinical and economic need (customer pain)
- How it is addressed today
- What gaps exist
- The primary and secondary customers

THE PRODUCT OFFERING

This section articulates how the product works and how it is protected. This may take more than one slide but should not take more than two slides. The presenter should be able to articulate the current status of the technology and any outcomes of the concept that have already been demonstrated. The company must be able to present a long-term opportunity.

Looking at our toolbox again for guidance, a well-prepared intellectual property pyramid can quickly express how the product works and how it is protected. Frequently, start-up companies like to bring in and show a prototype. The presenter should think this through, as sometimes a prototype results in a tangential discussion that can result in the presenter not having enough time to get their message across.

As it relates to the long-term opportunity, we expressed in the business plan that sometimes the first product into the market is only one in a series of products. We also commented that sometimes the first product into the market does not address the biggest opportunity. If this is true, that rationale should be provided here.

Below are the contents of the PowerPoint slide:

- Near-term product/service offering
- How it works
- How is it protected?
- Current status of the technology
- Demonstrated outcomes of the concept
- The long-term opportunity

THE MARKET OPPORTUNITY

This section is oriented to contextualizing the product opportunity within a particular market. Again, Figure 11.1 in Chapter 11 is very helpful for framing this opportunity. Investors are looking to understand that the market is growing from two perspectives. First, they are interested in understanding the factors that increase overall market value. Looking at Figure 11.1, an increase in patient prevalence and incidents results in an increased diagnosis rate. A new product that addresses a gap results in a new procedure. Increasing the procedure rate raises the entire product category. Last, changes to units per procedure can also increase the value in the market. Investors are highly interested in factors that increase overall market value. You may have heard the phrase "a high tide raises all boats"; the factors of prevalence, incidence, diagnosis, procedure rate, and units per procedure raise the market value of all participants. As such, fighting competitive companies is secondary to raising the value of the entire market. This is much more attractive and less risky for investors. As such, the contents of the PowerPoint slide should also include the following:

- Patient profile, prevalence, incidence, treatment methods
- Market size: growth rates of industry and target market
- Creation of new/untapped market demand
- Description of external factors driving growth
- Is the competitive advantage sustainable?

THE COMPETITION

The main questions that an investor is looking to answer are: Is the company positioned to compete effectively? Can the company gain access to the sales channel? Who else has a product that is addressing the same customer need and are there other products under development? Visiting our toolbox, the macro and micro maps, Figures 14.1 and 14.2 in Chapter 14, easily express the industry. Additionally, by utilizing the *Entry Schedule Fact Book*, Figure 18.2 in Chapter 18, the start-up can supply the details. However, these maps are generally subledgers to this PowerPoint slide. The start-up should create some form of table that easily expresses the participants in the market and the positions that they hold. It is important to note that the positioning in this table should be aligned with the "rule of three," your intellectual property, regulatory, and reimbursement messages. The contents of the PowerPoint slide should include the following:

- Macro: What companies play in these segments?
- Micro: What product lines does each company play with?
- What competing technologies are under development?
- Is the company positioned to compete effectively?

THE BUSINESS MODEL

The business model discussion should include your marketing and sales distribution strategy. For example, do you plan to build a sales force or use a distributor? Will you be using a contract manufacturer or will you be manufacturing the product yourself? What are your anticipated market shares and revenues in the first 5 years? What are your pricing and cost of goods sold targets?

For many pre-seed and seed stage investments, it may be too early to have the details associated with your business model. In that case, benchmarks or targets are useful for demonstrating that the start-up has reasonable projections. For example, if the start-up has not gone past the prototype, they can express their cost targets and communicate how they align with that requirement. The contents of the PowerPoint slide should include the following:

- Market strategy
- Manufacturing, market, sales, distribution strategy
- Anticipated market share
- Anticipated revenues 1 year/5 years
- If too early, provide a sense of scale
- Price of product/cost of goods sold (COGS)

THE COMMERCIALIZATION PLAN

The commercialization plan is the summary of the details provided in the business plan. How does your intellectual property position set the company up to effectively compete? What is your initial target market and why did you select it? What are the adoption drivers in the industry? It is most likely that it will take two to three slides to summarize the business plan. The contents of the PowerPoint slide should include the following:

- Intellectual property
- Market selection rationale
- Adoption drivers
- Clinical trials
- Reimbursement
- Sales channel and pricing strategy

FUTURE MILESTONES

This section articulates the milestones associated with the next round of funding and beyond. It is frequently a simple timeline that is expressed by many as a Gantt chart. The timelines are generally by quarter for the first 12 months, in half years for months 13 through 24, and annually for years 3 to 5. Each column summarizes the expenses associated with those time periods. Below that is a cumulative total so that investors can track the total equity into the company. Below that is a running total of cash available that includes when funding is raised and how it is spent.

The value of this approach is that it easily helps people identify how much the company needs to spend between fundable milestones. It allows a company to project how many months they expect to take to raise the new capital. The company must reserve cash for the time frame between the fundable milestones in the time that it takes to raise the new capital. Frequently, many start-ups miss this point and assume they can raise capital within 30 to 90 days. A fund-raise typically takes 6 months to a year. The contents of the PowerPoint slide were discussed in Chapter 18 and are provided below:

- Fundable milestone
- Clinical/regulatory
- Product launches
- Fund-raising timelines

THE MANAGEMENT TEAM

As discussed in the business planning section, the management team consists of the management board, the board of directors, and the scientific advisory board. It may take a few brief PowerPoint slides to detail the management team. Returning to Figure 8.1 in Chapter 8, we are reminded that the goal of the management team is to build out the network, leading to a liquidity event. In that spirit, revisiting Chapter 13, where we discussed that the gathering of domain-experienced personnel reduces risk, the highlights of those thoughts should be expressed in this slide. The content of the PowerPoint slide should include the following:

- Management team
- Scientific advisory board
- Board of directors

FUNDING NEEDS

The goal of this section is to detail the money raised to date. If there is non-dilutive funding such as an SBIR grant, it should be highlighted as it offers an independent third-party validation point. Many companies will create a PowerPoint slide such as the one used for future milestones and continue to build it out in this section.

This section should also discuss how investors will realize their profits and achieve an exit. Typically, companies will provide comparatives of recent exits. The company would be wise to uncover not only the most outrageous exits in the marketplace but also include realistic ones. In Chapter 17, we discussed funding milestones; it may be valuable and bring credibility to the conversation if the start-up can express its expected value at its next fundable milestone and compare it to that of the market. The content of the PowerPoint slides should include the following:

- Money raised to date
- Money currently being sought/use of funds
- Anticipated future funding needs
- How will investor realize profit/exit strategy

THE ACQUIRER'S NEEDS

The slide should be brief, yet express why an acquirer needs to purchase your company. How will you bring value to their business? Why can't the acquirer do it themselves or buy it from someone else?

OPPORTUNITY SUMMARY

The conclusion of the presentation should reiterate the venture concept and highlight the goals, the management team's talent, and expected returns. If the company will require corporate and venture capital, they should express the distance to the fundable milestone necessary to attract them. In that spirit, the company should identify the likely sources of follow-up funding, as frequently the fundable milestone to attract corporate and venture capital is a greater distance away than the next finance around. With this knowledge, the company should identify where they believe their follow-up funding sources are and express their ability to get to them. The content of these PowerPoint slides should include the following:

- How fundable is the company in the venture capital community?
- What are the likely sources of follow-up funding?
- Does the company have a clear understanding for finding follow-up funding?

DISCUSSION QUESTIONS:

1. Investors are looking to understand that the market is growing from two perspectives. Explain these two perspectives
2. What is the commercialization plan?
3. What is the business model, and how does it different from a commercialization plan?
4. What is the value of future milestones to investors?
5. What are the topics that should be discussed under funding needs?

NOTES

1 *Wikipedia*, s.v., "Abdominal Aortic Aneurysm," http://en.wikipedia.org/wiki/Abdominal_aortic_aneurysm (accessed March 15, 2014).

2 D.J. Katz, J.C. Stanley, G.B. Zelenock, Operative Mortality Rates for Intact and Ruptured AAA in Michigan: An 11-Year Statewide Experience, *Journal of Vascular Surgery*, May 19, 1994, *PubMed*, http://www.ncbi.nlm.nih.gov/pubmed/8170034.

20 Continuously Improve Your Message with the Plan-Do-Check-Act Cycle

As we review the four major parts of this book, we can conclude the following:

- Innovation is a process of connected steps.
- Investment must be connected to exit.
- Start-ups must align with industry norms.
- All constituencies must be told a compelling story, and that story is tested in the "act" moment of the Plan-Do-Check-Act (PDCA) cycle.

A start-up's interactions with customers, acquirers, and investors will provide information that should be immediately taken into a "check-act" moment in the PDCA cycle.

Angel, corporate, and venture capitalists predominantly invest in commercialization and have little interest in funding research: they invest in product development. Start-ups initiated by academics sometimes struggle with this understanding. The value of negative feedback is always an opportunity if one maintains the proper PDCA mindset.

For example, a health care IT company founded by an academic was targeting an attractive market. The academic CEO's projections to get to market were well within standard and he had received significant grants that de-risked the company from a value perspective. The grants also provided an independent assessment on the importance of the technology. A valued market, non-dilutive funding, and an independent assessment on the importance of the technology seems like a wonderful place to start.

After several months of failed attempts to raise capital, the author saw the academic CEO present at an angel network gathering. The presentation went fairly well until the funds slide was displayed. The academic CEO stated that 20% of the funds raised were going to be used on research for one specific aspect of the technology. The subsequent Q&A session offered a meaningful PDCA moment for the CEO, which he unfortunately never took to heart. The issue was that these angels did not want to invest in academic research, which was confirmed after the presentation. There was a post meeting social opportunity and time for discussions. The author had a moment to speak to some of the angels that he saw asking

DOI: 10.1201/9780367533052-20

questions during the Q&A session. He asked them what they thought about the opportunity. Although each angel verbalized the issue in his or her own way, the theme was that it was an attractive marketplace but the project was still a research project. The fundamental message was that his company was still perceived as an academic project.

The reality of the situation was that out of $1 million of funding that was requested, only about $150,000 was needed to complete the research aspect of the project. The issue was not research, it was *development*, and pointing out that subtle difference properly could have resulted in better alignment with investor language going forward. The author followed up with the academic CEO who refused to listen to the feedback and years later the technology still remains undeveloped. This CEO missed an opportunity to use the PDCA cycle to listen to the feedback and adjust his tone. Tone is the important lesson from this story as it was not the investment concept that was of concern to investors; it was the academic tone and focus with which it was presented. Because the CEO did not know his audience and was unwilling to internalize their feedback, he missed an opportunity to gain their support.

In another example, a promising pharmaceutical technology was making rapid scientific and clinical advancements. The company was founded by an industry thought leader who had unprecedented access to acquiring company management. As the company continued their advancement toward the clinic, the acquirers communicated the type of due diligence details that they were going to need to make an acquisition in the long-term. The CEO was extremely excited by the willingness of the acquirers to provide feedback. Upon reviewing their feedback, he recognized that his spending focus had not been on the supporting documentation necessary for a long-term sale of the company. Unlike our first story, the CEO added additional monies to his next fund-raise to hire the talent necessary to prepare him for an early exit. As of this writing, excitement and momentum continue to be afforded to this company whose CEO had the foresight to adjust to feedback from the PDCA cycle. What the CEO did not know was that the future acquirer's excitement, combined with the reputation of the thought leader, resulted in a venture capital firm being interested in investing early. Upon completion of their due diligence, a venture capital firm proceeded to invest because they felt that the design systems and the clinical systems were documented to the point that the venture firm felt the company could submit documentation to the FDA early. The benefits of a PDCA cycle are hard to predict. This CEO got a hint that his documentation was most likely not at an advanced state. One could argue that the CEO could have delayed the investment. However, the CEO ultimately benefited because his treatment of the situation impressed the acquirers. As acquirers have relationships with firms they acquire from (in this case a venture capitalist), the venture capitalist decided to take a look at the investment. Although the investment was normally a little early for this venture capital firm, the reputation and interest of the acquirer warranted their diligence. As the CEO was prepared for due diligence, he increased the enthusiasm of the venture capital firm as they determined that he was indeed an advanced property. Being able to invest in an advanced property early that has been de-risked by grants and expression of interest by the acquirers made all the difference for this company. The story demonstrates that one never knows the benefit that the check and adjust cycle can provide a company.

Start-ups must understand that the goal of investment pitch is to get another meeting. A 30-minute investment pitch rarely results in immediate funding. In fact, the "rule of three" would tell us that our goal would be to get to a 3-hour meeting to more deeply explore the opportunity.

Another aspect of an investment pitch is to get to know the other party. The start-up must appreciate that they should not always take money from those that are willing to invest in them. There are demanding and difficult investors in the universe that can consume a great deal of management's time. Although money is scarce, so is management's time.

If this concept is truly understood, the CEO approaches fundraising no differently than a sales professional approaches the sales pipeline. The fund-raiser will constantly have individuals that he is targeting, individuals that he is processing through the "rule of three" and individuals that he is closing.

Like a sales professional, the fund-raiser should have a balanced pipeline, as there is never certainty until the money is wired into the company's bank account. Every conversation and investment pitch is another opportunity to discover obstacles and impediments. The character and success of an organization grows as they use feedback to create a PDCA cycle and improve their approach. As they say—feedback is a gift. As the late Roberto Bolano stated, "If you're going to say what you want to say, you're going to hear what you don't want to hear." The difference between success and failure may be as simple as how you react to, and what actions you take in response to, what you did not want to hear.

DISCUSSION QUESTIONS:

1. What could be concluded after reviewing the four major parts of this book?
2. Discuss the value of negative feedback as an opportunity if one maintains the proper PDCA mindset.

Index

Note: *Italicized* page numbers refer to figures, **bold** page numbers refer to tables

B

C

D

E

F

J

K

L

S

T

U

V

W

Y

Z

Printed in the United States
by Baker & Taylor Publisher Services

Printed in the United States
by Baker & Taylor Publisher Services